D1367111

GOING WIRELESS

A CRITICAL EXPLORATION
OF WIRELESS AND MOBILE TECHNOLOGIES
FOR COMPOSITION TEACHERS AND RESEARCHERS

NEW DIMENSIONS IN COMPUTERS AND COMPOSITION
Gail E. Hawisher and Cynthia L. Selfe, editors

GOING WIRELESS

A CRITICAL EXPLORATION
OF WIRELESS AND MOBILE TECHNOLOGIES
FOR COMPOSITION TEACHERS AND RESEARCHERS

Edited by

Amy C. Kimme Hea
University of Arizona

HAMPTON PRESS, INC.
CRESSKILL, NEW JERSEY

Printed in the United States of America

Library of Congress Cataloging-in-Publication Data

Going wireless : a critical exploration of wireless and mobile technologies for composition teachers and researchers / edited by Amy C. Kimme Hea.
    p. cm.
Includes bibliographical references and index.
ISBN 978-1-57273-780-8 (hardbound) -- ISBN 978-1-57273-781-5 (paperbound)
  1. Telecommunication in education. 2. English language--Rhetoric--Study and teaching (Higher) 3. Multimedia communications. 4. Students--Information services. 5. Wireless communication systems. 6. Mobile communication systems.
I. Kimme Hea, Amy C.
  LB1044.84.G65 2009
  808'.042078--dc22

                        2008047696

Hampton Press, Inc.
23 Broadway
Cresskill, NJ 07626

In memory of my parents

CONTENTS

vii

PART III CUTTING THE CORD: STORIES ON WIRELESS TEACHING AND LEARNING IN THE COMPOSITION CLASSROOM

PART III TEACHING AND LEARNING IN MOTION: MOBILITY AND PEDAGOGIES OF SPACE

PART IV TEACHING AND RESEARCH IN MY POCKET: MOBILE GADGETS AND PORTABLE PRACTICES

APPENDIX

INTRODUCTION

Amy C. Kimme Hea

Integration of wireless and mobile technologies on college and university campuses has steadily risen over the past seven years. From Drexel University's claims to the first all wireless campus to Carnegie Mellon's "Wireless Andrew" research project, college and universities are moving away from traditionally wired systems toward wireless libraries, classrooms, dorms, and entire campuses. The 2002 Gartner DataQuest's campus computer survey reports that "70 percent of U.S. college campuses had some local area wireless network coverage, while 10 percent had full campus coverage" (Akin, 2003, p. 91). Mobile technologies such as cell phones and Personal Data Assistants (PDAs) continue to increase in use. Last year alone, 84 million cell phones with digital camera capabilities were shipped to users (Stone, 2004, p. 52). In addition, Comscore Networks claims that "10 million Americans surf [the Web] from cell phones and PDAs" (Ellison, 2003, p. 64). In the fall of 2001, the University of South Dakota required all entering students to purchase a PALM PDA (Akin, 2003, p. 92). And Duke University launched its APPLE iPod initiative in Fall 2004, providing all entering first-year students with the device. Experts in mobile technologies predict that handheld devices such as PDAs will become as ubiquitous in workplaces and college campuses as the desktop computer (Chen, 1999; Weiser, 1998).

This migration to wireless and mobile technologies means a shift in the pedagogical and curricular spaces typically reserved for writing instruction. First-year composition still represents one of the few standard require- ments for most college and university students, and more upper-division and writing-intensive courses are being offered across institutions. Increasingly, writing instructors are teaching such classes in wireless labs and engaging new mobile technologies with their students. These instruc- tors, then, are responsible for teaching not only rhetoric and composition but also technology. Similarly, Writing Program Administrators (WPAs) must invest in technology maintenance, supervision, and teacher develop- ment. Further integrating technology into the writing classroom now means we must better understand the historical, cultural, and instruction- al impacts of wireless classrooms and mobile technology use. Cited as a less expensive alternative to the wired classroom (Green, 2004), wireless laptops—either student-owned or cart-stored and classroom assigned—are being used in many of our first-year and advanced composition courses. Long used in medical and foreign language instruction, PDAs also are pro- viding new possibilities for field research and mobile, "just-in-time" instruc- tion. Minicomputers, Global Positioning Systems (GPS), cell phones, BlackBerrys, and other wireless and mobile technologies are making their mark on the curricular and social lives of educators and students. Wi-Fi net- works and Bluetooth technologies have encouraged students and teachers to engage in the "anywhere, anytime" educational practices touted among Information Technology (IT) specialists (Galambos & Abrahamson, 2002). All of these practices inform the roles of students, teachers, administrators, and scholars, and *Going Wireless* speaks to these shifting responsibilities.

SITUATING WIRELESS AND MOBILE TECHNOLOGIES IN RHETORIC AND COMPOSITION

Students entering the academy already live in a culture of wireless and mobile technologies. From cell phones to iPods, college students are often the target markets for such devices. Their consumer choice is only one rea- son this population exerts a powerful influence on developing technologies. A more profound rationale for marketing to college students is their future status as workers who will rely on wireless laptops, "just-in-time" invento- ry controls, BlackBerrys, cell phones, and other such technologies. In fact, college students' shift from loyal individual customers to managers and industry leaders also makes them appealing test markets for emerging technologies. John Kelly, Vice President of Business Development at 5G Wireless, explains to his business and IT colleagues that they should take higher education more seriously, not because we in rhetoric and composition offer a broad, critical perspective on technology and learning, but because pursuing higher education markets offers "a potential enterprise customer that provides industry with a viable real world example" (McGinty, 2003, Bottom Line 101, ¶ 8). Although we educators may want, and in some cases may even need, relationships with such corporate stakeholders, we also must strive for equity within those relationships. We need to understand the implications of corporate arguments about our classrooms because they affect the design and implementation of mobile and wireless technologies on our campuses as well as serve to constrain student and teacher subject positions. Randy Levine, Executive Director of Higher Education Wireless Access Consortium (HEWAC), asserts to other industry leaders that outfitting college campuses with wireless technology "will come in handy in the next sales call. I can't imagine a better opportunity to go to a Fortune 500 company and say we did implementation of 25,000-30,000 people" (McGinty, 2003, Bottom Line 101, ¶ 9). To hold any influence at all, to have any voice in the implementation of wireless and mobile technologies in our curricula, educators and researchers in rhetoric and composition must become better informed about the far-reaching discourses and practices of wireless and mobile technologies.

With this aim, *Going Wireless* is the first edited collection on wireless and mobile technologies in our field. *Going Wireless* contributors offer rhetoric and composition teachers, scholars, and administrators a range of practical and theoretical insights on wireless and mobile technologies. This collection seeks to serve as a resource for theoretical explorations on wireless and mobile technology use as it relates to computer and composition teaching and research and to act as a reference for the members of the rhetoric and composition community charged with the responsibilities of integrating, supervising, and evaluating wireless and mobile technologies. *Going Wireless* explores wireless and mobile technologies, raising practical and theoretical problematics.

CHAPTER OVERVIEWS

Going Wireless is organized into five major sections and an appendix of key mobile and wireless technology terms. In each section, authors represent a range of perspectives as they articulate the roles of students, teachers, administrators, and researchers working with and through mobile and wireless devices. *Going Wireless* provides readers with ways to understand the influence of wireless and mobile technologies by critiquing the corporate and IT perspectives that inform mobile and wireless development and integration and seeking out new tropes for learning, teaching, and researching. Drawing upon on interviews and surveys, rhetorical analyses, and theoretical explorations of wireless and mobile devices, *Going Wireless* authors enact a range of methodologies to make claims for reflective approaches to research and teaching with wireless and mobile devices. The contributors share their perspectives on the impact of these technological shifts and situate their experiences in relationship to rhetoric and composition as a field. More than assuming that these technologies are merely tools, authors in this collection argue for complex articulations of histories, deployments, integrations, and social factors affected by and affecting mobile and wireless technologies.

This collection offers more than an early-adopter perspective on emerging technologies. The growth of computers and composition, rhetorics of technology, and professional and technical communication frame the authors' views and inform critical positions expressed in the project. This critical stance may best be characterized by Donna Haraway's oft-cited position in her discussion of cyborg imagery. Haraway (1985) deploys the cyborg and other hybrid tropes to challenge the reductivism of universal, totalizing theories of technology and to embrace the partiality of our connections with our sense of self, others, and technology. *Going Wireless* chapters demonstrate the partiality of perspective that Haraway's work advocates—wireless and mobile technologies are articulations in process, not determinations that leave no room for agency. Further, this collection pursues Gail E. Hawisher and Cynthia L. Selfe's (1991) and Cynthia L. Selfe's (1999) hope that rhetoric and composition students and instructors exercise their agency to become critical users, producers, and critics of technology.

Going Wireless also claims that such agency is related to technological literacy. Anne Wysocki and Johndan Johnson-Eilola (1999) remind us that if we equate "technological literacy" with mere skills, we can mistakenly reenact early literacy movements that also constructed reading and writing instrumentally. The instrumental approach arguably positions literacy outside of political, ideological, and material fields. It is worth stating Wysocki and Johnson-Eilola's position that we must "understand ourselves as active participants in how information gets 'rearranged, juggled, and experimented with' to make the reality of different cultures. This involves, of course, understanding ourselves within the making and changing" (p. 366). For his part, Stuart A. Selber (2004) acknowledges the danger of failing to adopt a postcritical stance on technology, arguing that we in rhetoric and composition may unwittingly situate technological development outside our purview (p. 13). Striving for the contextual, critical, and historical perspective that Selber advocates, *Going Wireless* posits that technological literacy is bound in relations of politics, power, and agency, and that rhetoric and composition scholars, teachers, and administrators are already intimately connected with the cultural practices of wireless and mobile technologies.

SECTION 1: REFIGURING WRITING, TEACHING, AND LEARNING THROUGH WIRELESS AND MOBILE TECHNOLOGIES

In the Refiguring Writing, Teaching, and Learning through Wireless and Mobile Technologies section, well-respected scholars and educators in computers and composition consider the ways in which mobile and wireless devices refigure issues of power, knowledge, writing, and teaching. With the emergence of Short Message System (SMS) authoring and other related genres, Johndan Johnson-Eilola and Stuart A. Selber discuss what counts as "composition" in their piece, "The Changing Shapes of Writing: Rhetoric, New Media, and Composition." Johnson-Eilola and Selber offer a simple, flexible framework, which they call C3T—context, change, content, and tools—to approach communication situations where "traditional writing" is only one of many possibilities. Similarly, Teddi Fishman and Kathleen Yancey explore wireless technologies in their work, "Learning Unplugged." Making their point that wireless changes not just the *how* but also the *what* of teaching composition, Fishman and Yancey ask us scholars, teachers, and administrators to consider our criteria for the selection of pedagogical materials, intellectual property and assessment issues, and assumptions about classroom sites. Johnson-Eilola and Selber and Fishman and Yancey raise questions about the shape of writing and the cultural assumptions about composing in a range of media, from the small and portable to the classroom-integrated. These scholars ask us to question and revise not just our roles as teachers, scholars, and administrators but also our definitions of what it means to "compose" in the shift from wired to wireless technologies.

SECTION 2: EXAMINING TEACHER AND STUDENT SUBJECTIVITIES IN THE AGE OF WIRELESS AND MOBILE TECHNOLOGIES

In the Examining Teacher and Student Subjectivities in the Age of Wireless and Mobile Technologies section, the book turns directly to the ways in which wireless and mobile technologies complicate the roles of teachers and students. As Karla Saari Kitalong argues in her piece, "'Whole New Breed of Student Out There': Wireless Technology Ads and Teacher Identity," teachers' roles are shaped by the discourses of wireless technologies in the academic realm. Kitalong reviews several ads in the widely distributed pedagogical magazines, *Syllabus, Campus Technology,* and *Edutopia.* Arguing that these ads constrain teacher identity potential with respect to the educational use of wireless technology, Kitalong offers "intervention strategies" teachers can integrate to extend their subjectivities. Ryan M. Moeller also attends to cultural constructions of wireless technologies and their relationship to teacher positions in his essay, "ReWriting Wi-Fi: Changing Wi-Fi Proliferation through Writing Instruction." With the advent of mobile and wireless technologies, Moeller explores the difficult issues of surveillance and regulation. Rather than positing a determinist view, Moeller argues that students and educators can revise dominant discourses about these technologies through both qualitative and sociological Wi-Fi research and their own computer use. In "Reterritorialized Flows: Critically Considering the Roles of Students in Wireless Pedagogies," Melinda Turnley focuses on the problematic of access and student subjectivities. Turnley offers compositionists a complicated analysis and articulation of student positions as related to wireless and mobile technologies. Drawing upon noted critical theorists Gilles Deleuze and Félix Guattari, Turnley employs their concepts of deterritorialization and reterritorialization to theorize a productive position on mobility and its potential for student agency. These scholars adeptly argue that technology initiatives are always already informed by larger cultural conversations. These narratives impact us and our students in profound ways. Whether or not we have yet fully integrated mobile and wireless in our pedagogy, Kitalong, Moeller, and Turnley remind us that we are still part of the broader cultural conversation and that we can assert agency in relationship to the discourses and practices of wireless and mobile technologies.

SECTION 3: CUTTING THE CORD: STORIES ON WIRELESS TEACHING AND LEARNING IN THE COMPOSITION CLASSROOM

The Cutting the Cord: Stories on Wireless Teaching and Learning in the Composition Classroom section attends to research projects examining the integration of wireless and mobile devices across educational institutions. Interrogating teacher, student, and administrator roles, these chapters provide a glimpse into classroom and administrative praxes. Logistics, security, privacy, and cultural assumptions about technology all come to the fore as these scholars describe their work with mobile and wireless learning and teaching. In "From Desktop to Laptop: Making Transitions to Wireless Learning in Writing Classrooms," Will Hochman and Michael Palmquist revisit the groundbreaking *Transitions* study and discuss the integration of wireless computing in the writing classroom. Hochman and Palmquist report on the results of a fifteen classroom study of Southern Connecticut State University's wireless laptop composition instruction. Much like the participants in the *Transitions* study, these instructors faced profound pedagogical, not just technological, issues. Hochman and Palmquist posit that we must better understand the ways laptop classrooms reconfigure our teaching roles and prepare composition instructors to teach in these reconfigured spaces. Instructor integration of wireless technology is also at the center of Kevin Brooks' chapter, "Changing the Ground of Graduate Education: Wireless Laptops Bring Stability, Not Mobility, to Graduate Teaching Assistants." Brooks discusses the development of North Dakota State University's English department wireless laptop initiative for Graduate Teaching Assistants (GTAs) in the MA Program. Begun in the fall of 2004, this initiative provided each GTA with a wireless laptop, software package, and connectivity to three wireless access points in the GTA office spaces. Drawing on Marshal McLuhan and Eric McLuhan's notion of figure/ground, Brooks argues that those implementing wireless technologies need to pay attention to the whole ground of academic computing and not be overly distracted by the newest technological figure. Loel Kim, Susan L. Popham, Emily A. Thrush, Joseph G. Jones, and Donna J. Daulton provide us with a detailed investigation of mobile and wireless technology pedagogy in their essay, "A Profile of Students Using Wireless Technologies in a First-Year Learning Community." Based upon their fall semester 2004 research project with 17 first-year composition students who were members of a health-care learning community, Kim, Popham, Thrush, Jones, and Daulton create a profile of these students' personal and academic uses of wireless and mobile technologies. Rather than arguing for a top-down model of technology integration, Kim, Popham, Thrush, Jones, and Daulton seek to develop pedagogies and formulate policies drawn from student use, not teacher or administrator desires. They note the significance of economic and gender factors in their own research and prompt composition teachers and scholars to consider these issues in our own wireless classrooms. In "Security and Privacy in the Wireless Classroom," Mya Poe and Simson Garfinkel draw attention to security and privacy issues with wireless networks. Their chapter explains specific security and privacy alerts and demonstrates ways in which security affects teachers and students in wireless writing classrooms. Poe and Garfinkel assert that writing teachers and WPAs should understand security and privacy issues and educate students about the social implications of writing in wireless environments. All of these authors make praxis the center of their research and offer us initial insights into the complexities of integrating mobile and wireless technologies in our classrooms, curricula, and institutions.

SECTION 4: TEACHING AND LEARNING IN MOTION: MOBILITY AND PEDAGOGIES OF SPACE

The Teaching and Learning in Motion: Mobility and Pedagogies of Space section closely examines the potentials and constraints of mobility and the various articulations of mobile life. Drawing upon the ways in which mobile technologies are integrated into work, schools, museums, and social life, contributors in this section frame issues of ubiquity and mobility. In "Perpetual Contact: Articulating the Anywhere, Anytime Pedagogical Model of Mobile Composing," Amy C. Kimme Hea argues for a close examination of the trope of ubiquity associated with wireless laptop initiatives. Drawing upon ubiquitous computing (ubicom) and non-place theories, Kimme Hea articulates the anywhere, anytime development models of Tucson's Empire High laptop project and MIT's One Laptop Per Child (OLPC) program. She contends that compositionists must understand the potential effects of wireless discourses and practices for literacy and student and teacher agency so that a range of critical practices for wireless and mobile technologies can be reflected upon and enacted. In "Writing in the Wild: A Paradigm for Mobile Composition," Olin Bjork and John Pedro Schwartz also explore the effects of mobile technologies on composition.

Bjork and Schwartz turn to these new contexts for writing and encourage composition teachers and scholars to situate writing as a field process, one in which students explore various sites of rhetorical activity using mobile technologies to write and publish on location. Drawing upon the model of museums and new media mediations of museum exhibits, Bjork and Schwartz argue for a reinvigorated sense of place, space, and interaction in composition. This place-based model can be achieved through mobile technologies, particularly the mobile Blog (moBlog). Nicole R. Brown takes up the issue of mobile computing in her essay, "Metaphors of Mobility: Emerging Spaces for Rhetorical Reflection and Communication." Brown echoes Bjork and Schwartz's call for place-based composition as she examines a range of civic and social rhetorical situations that employ wireless networks and mobile technologies. For Brown, the metaphors of graffiti and public art are productive tropes for the public, resistant deployments of mobile technologies and student agency. Enacting pedagogies of mobile composition, instructors and students can position writing as public discourse. Through spatial theory, Kimme Hea, Bjork and Schwartz, and Brown frame mobility and its constraints and potentials. These scholars argue for the situatedness of our work as writing teachers, scholars, and administrators.

SECTION 5: TEACHING AND RESEARCH IN MY POCKET: MOBILE GADGETS AND PORTABLE PRACTICES

In the Teaching and Research in My Pocket: Mobile Gadgets and Portable Practices section, scholars turn to specific technological manifestations offering compositionists and rhetoricians a critical perspective on the impact of PDAs, internet radio, and the iPod. This last section emphasizes the diversity of mobile and wireless technologies and the broad influence they have in many domains. Clay Spinuzzi's chapter, "The Genie's Out of the Bottle: Leveraging Mobile and Wireless Technologies in Qualitative Research," provides rhetoric and composition scholars insight into using a PDA for research. Explaining the applicability of mobile and wireless computing for field notes, interviews, artifact analysis, and project management, Spinuzzi posits that new tools not only refigure our old research tools but also lead to a reexamination of qualitative research in our field. Steering us toward the medium of internet radio, Dene Grigar and John F. Barber theorize aurality and its impact on our field in their essay, "Winged

Words: On the Theory and Use of Internet Radio." Grigar and Barber established their own internet radio site Nouspace in 2003. Aural theories come to bear on Grigar and Barber's analyses of oral practices and analog broadcasts. Arguing that internet radio broadcasts have pedagogical and theoretical impact on our field, Grigar and Barber discuss the transformative aspects of the aural environment of internet radio. Then, Beth Martin and Lisa Meloncon Posner draw attention to the iPod in their essay, "Dancing with the iPod: Exploring the New Wireless Landscape of Composition Studies." Using cultural geographer Donald Meinig's notions of successive landscapes, Martin and Meloncon Posner apply his work to mobility. Martin and Meloncon Posner describe the features of the iPod and distinguish its uses and cultural capital of "cool," arguing that the iPod is a means to access a landscape of information and its manipulation. Spinuzzi, Grigar and Barber, and Martin and Meloncon Posner closely examine specific technologies and the ways these portable devices and transmissions influence researchers, teachers, and students. As mobile and wireless technologies are further integrated into our own and students' pedagogical and personal lives, rhetoric and composition scholars must consider the broader impact on our praxes and the ways we live and learn.

To extend the work of these chapters, David Menchaca explains many of the technologies, their histories, and permutations. Giving us a way of engaging the technologies in our research and instruction, Menchaca's chapter, "Terms for Going Wireless: An Account of Wireless and Mobile Technology for Teachers," is not just a dictionary of terms. Rather, Menchaca provides the historical context of wireless and mobile technologies through the lens of rhetoric and composition practice and theory, making his contribution broadly applicable to teachers, scholars, and administrators in our field.

Going Wireless articulates a far-reaching, multivocal dialogue on mobile and wireless technologies as cultural, technological, and historical manifestations that inform and are informed by our praxes as teachers, researchers, and administrators. Rather than assuming these technologies are neutral tools, all the contributors in *Going Wireless* take on difficult issues of integration, use, and development. Instead of providing answers, these authors begin a dialogue on the range of technologies available to us and our students. Neither celebratory nor reactionary, *Going Wireless* creates a space for critical reflection in the hopes of engendering a much-needed exploration on the discourses and practices of mobile and wireless teaching and learning.

REFERENCES

Akin, Jim. (2003, Fall). Unwiring at school. *PC Magazine*, pp. 91-94.

Chen, Anne. (1999, September). Handhelds on deck. *InfoWeek*, pp. 67-72.

Ellison, Craig. (2003, Fall). Unwiring your home. *PC Magazine*, pp. 64-66, 68, 70-75.

Galambos, Louis, & Abrahamson, Eric John. (2002). *Anytime, anywhere: Entrepreneurship and creation of a wireless world*. New York: Cambridge University Press.

Green, Kenneth C. (2004, May 1). Beginning the third decade. *Syllabus: Technology for Higher Education*. Retrieved May 20, 2004, from http://www.syllabus.com/print.asp?ID = 9364

Haraway, Donna. (1985). A manifesto for cyborgs: Science, technology, and socialist feminism in the 1980s. *Socialist Review, 80*, 65-105.

Hawisher, Gail E., & Selfe, Cynthia L. (1991). The rhetoric of technology and the electronic writing class. *College Composition and Communication, 42*, 55-65.

McGinty, Meg. (2003). Educating wireless. Retrieved May 20, 2004, from http://www.5gwireless.com/company/EducatingWirelessArticle.htm

Selber, Stuart A. (2004). *Multiliteracies for a digital age*. Carbondale: Southern Illinois University Press.

Selfe, Cynthia L. (1999). Technology and literacy: A story about the perils of not paying attention. *College Composition and Communication, 50*, 411-436.

Stone, Brad. (2004, June 7). Your next computer. *Newsweek*, pp. 51-54.

Weiser, Mark. (1998). The future of ubiquitous computing on campus. *Communications of the ACM, 41(1)*, 41-42.

Wysocki, Anne Frances, & Johnson-Eilola, Johndan. (1999). Blinded by the letter: Why are we using literacy as a metaphor for everything else? In Gail E. Hawisher & Cynthia L. Selfe (Eds.), *Passions, pedagogies, and 21st century technologies* (pp. 349-368). Logan: Utah State University Press.

PART I

Refiguring Writing, Teaching, and Learning through Wireless and Mobile Technologies

1

THE CHANGING SHAPES OF WRITING

RHETORIC, NEW MEDIA, AND COMPOSITION

Johndan Johnson-Eilola

Stuart A. Selber

Technology now offers new possibilities for reading and writing. This means that the repertoire is expanding. Older forms, such as the book, are not defunct, but learners need to acquire new literacy skills.

—Futures: Meeting the Challenge (2005, p. 17)

This chapter encourages the expansion of what counts as "composition" by considering the authoring of cell phone text messages and related genres, which we see as an appropriate and worthwhile assignment for students. We offer a simple, flexible framework for approaching communication situations in which traditional "writing" is only one of many possibilities. The framework involves context, change, content, and tools (abbreviated as C3T), four aspects that help to situate messaging (and other communication) practices in broad rhetorical terms.

THE SLOW DEATH OF WRITING

Many of us are familiar with messaging technologies primarily as they occur in our classrooms: students using INSTANT MESSENGER (IM) or cell phone text messaging capabilities to chat with friends. The opinion of many academics about the intellectual worth of such technologies is captured in this sign (see Figure 1.1), which was posted above computers in a campus library we visited while attending a professional conference.

Often, particularly when they take place during class sessions, these exchanges bother us. Not only do they seem rude—this is, we think, *our* class time—they also seem frivolous, not worthy of careful study. Anyone, it seems, can send a text message—"real" writing is harder to learn.

Like many disruptions, these instances are also teachable moments, suggestions about new possibilities. For although "traditional" writing will remain, for the time being, a mainstay of the "writing" classroom, new forms of communication are increasingly influencing and sometimes displacing our treasured genres: the essay, the report, the short story. We don't want to argue that these new forms are inherently better or worse than the old ones. That argument misses the point: Those new forms are here and are increasingly popular. Recall that during the 1980s, many writing teachers argued about whether using a computer could improve writing instruction (Hawisher, 1988). Results, predictably for such a complex object, were mixed. And while we argued, the world moved on. So now, for most of us, writing is a computer-based activity, something we address as a normal aspect of writing rather than an innovation to be rigorously tested before admittance. Questions about whether or not new forms such as IM and Short Message Service (SMS) hurt writing are beside the point: They're already here—and our students are using them on a daily basis.

Figure 1.1. Sign Above Computers in a Campus Library[1]

So rather than ignore or deride these new forms, we'd like to suggest that writing teachers begin integrating them into their classrooms. Although this addition can be threatening to some traditional pedagogies that prioritize modes of writing or grammatical correctness, those systems are rapidly disintegrating under the complexities of contemporary life. In the framework we lay out here, traditional issues such as genre, grammar, and style are not removed, but placed within a broader, flexible context as specific instances or practices among many others. In one way of thinking, we are offering an extension of the process-based approach that is designed to deal with more complex writing contexts and changing technologies. Teachers are, in this framework, experts in rhetorical analysis and communication processes. As new situations arise, teachers help students to draw on their existing knowledge of communication, to analyze and plan communications, and—not incidentally—to discover and learn new techniques and technologies as the evolving situations demand. When faced with a novel situation, teachers should not expect to rely solely on their own existing knowledge, but rather they should be comfortable saying, "That's interesting. Let's figure out how to learn to respond to it, how to learn it, how to use it effectively." Experimentation and exploration are key tactics in the C3T approach.

This effort would have immediate benefits for the status and role of writing instruction. First, by recognizing the validity of these forms, compositionists acknowledge that communication is a broad and varied enterprise, one that surfaces in a surprisingly large number of areas in all of our lives, including our students'. Many people have bemoaned the decline of literacy in the age of the Internet (Birkerts, 1994; Sanders, 1994). But students write all the time, probably more than they did twenty years ago. It's just that we're only now starting to recognize what they do *as* writing—or more broadly, as important forms of communication. Some teachers, for example, are starting to teach web design, Weblog authoring, and synchronous discussion (chat sessions, MOO collaboration) as forms of writing. To these activities we would add wireless forms of writing such as SMS and IM. If we admit these new forms, we validate not only those new forms, but older ones as well.

Which brings us to a second benefit: If we worry that, for example, the abbreviations and misspellings of text messages are damaging students' abilities to write more traditional forms, we need to take a rhetorical approach. We need to demonstrate to students that conventions of text messaging are completely appropriate to some specific situations, whereas different conventions work more effectively in other situations. We should not reframe the old high-low culture divide, prizing more traditional forms of writing over newer ones simply *because* they are older (any more than we should elevate new forms of writing simply because they are newer).

Finally, a rhetorical stance helps to validate important knowledge that students already possess about how to communicate. Although we can continue to offer them important help in learning how to communicate what we think of as traditional forms of writing, we can also help them to build bridges between what they currently know and what we can help them come to know. We should not dismiss, for example, a student's ability to text message as an inferior form of writing: Effective text messaging requires sophisticated skills of understanding concrete rhetorical situations, analyzing audiences (and their goals and inclinations), and constructing concise, information-laden texts as part of an unfolding, dynamic social process with others. The fact that many of us do not possess these skills should not prevent us from appreciating their rhetorically rich nature.

In many cases, we've become so comfortable with our own traditional expertise that we have become dismissive of new forms of communication that make us feel like novices. If we can give up the notion that teachers must be masters of all terrain that the course covers, we can start to understand our expertise in different, more useful ways. We may even begin to learn *with* our students. Let us keep in mind that valuing new forms does not automatically demote other communication practices. Instead, it should encourage us to develop a broader framework for understanding how communication occurs, one that can capture knowledge and practices for writing text messages *and* essays, for exchanging peer-group critique comments *and* e-mail.

None of this, it should be clear, is radically new: Forms of writing have always undergone change and always met with resistance (for historical perspectives, see Eisenstein, 1979). But rather than fight running skirmishes over each new technology as it emerges, we want to step back and consider the notion that taking a broad rhetorical approach to communication—something we have long claimed to do, but frequently only in limited ways—can provide us with a framework that can adapt to new technologies as they are developed, taken up, and used by people in concrete situations.

The need for an open framework is crucial, if for no other reason than the fact that the pace of technological change does not appear to be slowing down. If we tie our teaching to specific forms and genres, our teaching will start showing its age within only a few years (or months), as new communication technologies come into use. Technologies and uses mutate, converge, diverge, and scatter. Consider the development of the telephone. In addition to supporting traditional, synchronous voice conversation, the modern cell phone hosts a surprising array of communication spaces: web browser, e-mail system, text-based chat, still and video camera, calendar, alarm clock, even remote control. In fact, some of our students use their cell phones as much for capturing events as for exchanging messages. Similar mutations have occurred in other technologies as well: Computers now act as videoconferencing systems, televisions, recording studios, telephones, and more. Futurists have long talked about "converging technologies," but in fact we're seeing a simultaneous convergence and divergence. The functions of individual technologies are now fluidly moving about, with any single communication device affording not only its traditional functions, but also those of many other technologies (see Figure 1.2).

As we said earlier, many of us see messaging technologies (both wired and wireless) only at the edges of our vision. But they're having a broad and growing impact on many groups of people. Messaging (frequently referred to as "texting" in the context of cell phones) is a frequent activity

Figure 1.2. Journalism Student Moblog at University of Georgia, Covering 2004 Election[2]

not only for students, but also in the military (McCarter, 2003), in the corporate world (Reid, 2004), and more. For example, a corporate communications consultant we know recounted the necessity of reminding Blackberry users in the financial industry of the rhetorically problematic issues involved with sending a short message such as "Sell!"

The framework we'll use below is almost laughably simple. It could be elaborated and complicated extensively, but at heart it consists of four aspects of writing: context, change, content, and tools. None of them are surprising, and all have been studied and taught by various people within the field. What is important is that they offer a relatively inclusive set of concerns that can drive decisions about communication, and they are open enough that they can adapt to varying situations and media. Indeed, we're attempting here to construct a framework that's open enough to absorb or articulate to other approaches to communication.

Dealing with computer technology in the writing classroom has occupied our field in primary ways for the past three decades (Baron, 1999; Heba, 1997; Kemp, 1992; Selfe, 1989; Yancey, 2003; among many others). Our goal is to extend this work by stepping back from a focus on specific technologies and focusing more on the overall rhetorical aspects of such approaches. We have, therefore, chosen to emphasize a project-based approach to learning that opens up the context for writing, reading, and using texts and requires students to learn about and choose specific technologies that work most effectively for the contexts in which they (and their readers) are working: printed text, e-mail, phone call, MACROMEDIA FLASH animation, digital video, web site, and more.

The aspects of this framework are listed in roughly the order that they would be considered in any writing situation. To some extent, the first two aspects (context and change) overlap because each is bound up with the other. They're listed separately primarily for analytical purposes.

Context: The situation(s) in which communication and living occur. Includes both micro and macro levels: from specific participants and pre-existing texts in a context to larger institutional, political, and social forces (among other things). Context may be (and typically is) multiple if the writer/designer and the reader/user are in very different circumstances, doubly so if anyone involved inhabits multiple contexts in relationship to their work. Roughly speaking, context is synchronic.

Change: As you might expect, this element deals with crucial aspects of a context or contexts that someone within them would like to change. This is the exigency for communication. Because change occurs over time, it is the diachronic aspect of communication. In theory, although not always in practice, change occurs within contexts in relationship to one or more communications. (It's possible, obviously, to discuss change in the absence of communication, but we don't want to cast our net that broadly. But in some writing situations, change may occur in spite of or alongside communications.)

Content: In communication theory terms, content is the information in the message. For our purposes, we're using the commonsense of the term: the words on the screen, the animated graphic demonstrating a procedure, the emoticon suggesting a joke. As suggested in the Context section, content frequently pre-exists the (artificially) isolated activity of authoring a specific communication: previous documents that are related to the document(s) currently being developed, resources on which a communicator might draw (libraries, databases, web sites, etc.).

Tools: Tools exist at the technological, concrete level: word processors, chat programs, SMS capabilities, and the like. We're slightly wary of the term "tools," because it suggests a relatively simplistic approach to understanding writing environments. And we place it last not because we think it's the simplest aspect—in some ways, it can be the most complicated (both functionally and ideologically). Tools can include specific technologies (Microsoft WORD shapes texts differently than NOTEPAD or DREAMWEAVER or AOL INSTANT MESSENGER) and conceptual practices (the C3T framework itself is a tool in this sense; so is genre, when understood as a semistable method for structuring content that occurs over time and space in routine ways in specific contexts). That is, we're both accepting and challenging the category in ways that we think expand the conceptual and functional potential of tools.

One key factor to keep in mind is that each of the aspects of C3T function for both writers and readers at all times. Writers are also always readers; the content created by one person becomes the content drawn on by others (or the same person) in subsequent communications; tools are used for both writing and reading texts (a printed book offering a different tool for reading than an editable Wiki page). This infinite extendibility means that although "writer" and "reader" are often useful terms, we might gain a lot by beginning to think of "participants" in communication, of "users" of texts in contexts.

Although initially people move through this framework in relatively linear order, after a first pass any but the most mundane situation will require an increasingly complicated series of recursions, revisitings, and revisions, as consideration of one aspect invariably influences other aspects. For example, in even the first run through the framework we might discover that thinking about contexts revealed the possible tools that might be used within the communication processes. And, as often happens in rich writing practices, working on the content of a specific communication may spark new ideas about desired changes in and across contexts.

One reason we've ordered the aspects linearly is to invert what has traditionally become the starting point for writing instruction: The ground rules for communication frequently assume certain genres, specific media. It is, we say, after all a writing class. But we're asking that you provisionally turn that idea on its head and assume that students will write in a particular genre only if that genre seems to be the best response to the contexts, desired changes, and content. Much of what we're discussing in this chapter grows out of our own work in project-based technical communication courses, in which students are frequently asked to design many different types of texts, depending on the emerging situation. As practicing technical communicators have amply demonstrated, complex, real-world solutions frequently require innovative thinking about media. When you have a hammer, as the saying goes, everything looks like a nail. (This realization is itself associated with the rise in things like text messaging, as users began to adapt new technologies to their own situations and discovered that a brief text message would be more effective than a memo, phone call, or e-mail.)

The open nature of our framework was designed to support project-based courses, because of the complexity inherent in such situations. Rather than organizing coursework around sets of skills or types of texts or specific tools, a project-based approach to communication requires an extremely flexible and open pedagogy, one driven by the changing and complex demands of realistic (or real-world) situations rather than predetermined, relatively static concerns. This approach can seem unsettling to teachers used to courses driven by a coverage model, but given the rapidly changing nature of communication in our culture and the need for communication skills to adapt to complex situations, the project-based approach is especially useful in helping students build a generalizable framework for approaching writing situations. In other words, a coverage model suggests that teachers can predetermine enough aspects of the writing situation to cover all contingencies. If this was ever true (a fact that we might dispute), it is no longer so.

If you haven't used a project-based structure for classes previously, we should warn you now that it will appear to be messy and extremely contingent, because there are fewer ground rules than exist for genre-driven approaches. And in many cases, as students (and everyone else, including teachers) experiment and learn new forms of media on the fly, the texts they design will frequently include missteps, inappropriate responses, and other failed gestures. That's perfectly fine; it's part of learning. Becoming comfortable with that contingency is itself a crucial aspect of the framework.

In the next section, we'll offer two examples to explore how the simple framework we've constructed can help students respond to communication situations. Both examples are fictional but typical within the contexts described. In the first, we'll take up a relatively simple, everyday situation: A student working on a project for a course needs to get information from another student working on his or her team. This example, though mundane, will help to illustrate instances in which new media such as wireless technologies are more effective within many contexts. The example will also demonstrate how students use their knowledge about genres, technologies, and situations to respond in effective ways to specific, unfolding situations. In the second, slightly more complex example, we examine how the framework can help students in an advanced, project-based course understand the demands of context and use an array of technologies to effect changes within that situation.

EXAMPLE 1: COMMUNICATING TO LOCATE KEY INFORMATION

In order to surface some key issues surrounding the use of wireless technologies and to demonstrate the process of using the development framework outlined, we'd like to consider a relatively trivial scenario.

Emily returns, empty handed, to the table she's staked out at the library. She's doing some initial research for a team project on Sunday afternoon. At this point in the project, her main responsibility is to assemble a brief review of existing research. During an earlier session scouring the Web for resources, she gathered some useful overview material but discovered that several reports referenced a landmark study that, as far as Emily can tell, is only available in print. She had accessed the library's web-based catalog on Friday morning and discovered that the library indeed has a copy of the book in its holdings. After she decided she would actually need to go to the library, she adds several additional resources that are available only in print.

At the library, though, she's hit a brick wall: Although Emily is able to find some of the resources she needs, the most important one isn't on the shelves when she looks for it. She connects to the library's online catalog and double-checks the call number for the book. The call number was right, but there's a new problem: Between Friday morning and Sunday afternoon, someone else has checked out the volume.

She looks at the screen for a minute or two, trying to decide her next step. She can submit a request to the library that the volume be recalled, but that could take a week or more. She can skip using this resource, but based on what she's seen so far, that would leave an important gap. She briefly considers faking it and covering the material in the book by cobbling together the paraphrases and quotes she can find in secondary sources. What she really wants to do is just borrow the book for an hour from whomever has it to photocopy one particularly relevant chapter—but the library, citing privacy issues, has a policy against giving out the names of people who have borrowed books. (A good policy, she admits.)

Then she realizes that it's likely one of her teammates has checked out the book. Given the specialized nature of the topic, it's unlikely (though not impossible) that anyone else would want the book. Although Emily's team has coordinated their work at a high level, Emily guesses that the material in the book might be useful for the portion of the report that someone else is writing. After mentally reviewing each team member's responsibilities, Emily decides that Vijay is the most likely candidate. The question now is how to contact Vijay. She runs quickly through her options: email, phone call, face-to-face, IM. Because her portion of the project is due tomorrow morning, she wants to resolve the issue quickly. She checks her Buddy List, but Vijay's away message says, "Out to lunch." That's not helpful.

She opens the window with her e-mail client in it and sends a quick message to Vijay:

> I'm at the library trying to find Dickinson's book on protocols, but it's checked out. Do you have it? Let me know ASAP.
> Thx.
> - E

After waiting a few minutes for a response, she thinks about calling his cell phone, but because she doesn't know whether or not Vijay is someplace where he can be interrupted by a call, she decides on the less-intrusive option of sending his phone a text message. She opens her cell phone's address book and selects Vijay's entry. She selects "Send SMS" from the options menu and writes a quick text message:

> do u have dickinson's book on protocols? can i borrow it for 1 hr?

She hits Send, and in about twenty seconds, a response comes back:

> yep. stop by my apt after 6.

Great. She sends a quick thank-you message to Vijay, and then returns to the pile of journals on her table to begin skimming them for additional information.

Applying the Framework

As we argued earlier, the context, change, content, and tools (C3T) model is flexible enough to apply productively to nearly any communication situation, without preset notions of genre, style, or communication technology. To some extent, many of us have already internalized key aspects of the framework and, for familiar situations, actually work through the processes without conscious effort. But analyzing Emily's activities and thought processes retrospectively can illustrate some important features of the framework and process (Table 1.1).

Context: This includes (at the very least) a person's immediate physical context as well as the broader social contexts in which they work. For Emily, the physical context of the library offers both constraints and opportunities. An obvious opportunity is access to print materials and space for laying out books, papers, and her computer while she's working. And access to the Internet via one of the library's wireless access points allows

TABLE 1.1 C3T CATEGORIES AND FEATURES FOR EXAMPLE 1

C3T CATEGORIES	FACTORS
CONTEXT	*Immediate*: Table at the library. Access to wireless Internet, space to lay out papers, books. Conventionally fairly quiet. *Broad*: Team project for college course. Teammates distributed physically at the moment (in unknown locations).
CHANGE	Construct useful and appropriate literature review for project. Locate information about missing key resource.
CONTENT	Broadly, a request to teammate about location of key resource. Adapted and changed based on tool used.
TOOLS	Numerous possibilities: phone call, SMS, IM, e-mail, F2F.

Emily to do quickly things such as send e-mail and consult the library's online catalog. Other features of the library such as the generally acknowledged prohibition on unnecessary noise are helpful in that they provide Emily a quiet space in which to work, but limiting in that the "quiet, please" atmosphere discourages her from using her cell phone to talk directly to Vijay.

The broader social context in which Emily works includes several key players: her teammates, the class in which the project assignment was made, and her campus community. Her connections within this social realm require her to consider the activities and responsibilities of her teammates and the expectations of her instructor (including issues such as the fact that the team's report needs to be structured in a conventional way for the discipline in which they're working).

Change: In nearly any situation, participants are working to bring about some sort of change. In this particular case, Emily is managing a whole host of change activities, loosely held together by the team's desire to earn a good grade for the project. In the example, the key change Emily deals with is the need to find out the location of an important book. The need for change, in concert with context, drives a recursive series of decisions, reflections, and actions that involve not only context but also content and tools.

Content: As we discussed above, content activities are often bound up tightly with tools activities. In this example, Emily's desire for change (*locate book*) in her particular context calls on her to take a more general need and articulate it in several different media. In each medium, the content of her communication is adapted to the specific affordances of the technology.

Tools: As Emily constructs the content of her various communications, she draws on her experiences with specific media in terms of genre expectations (even though she might not ever explicitly think, "E-mail obeys *these* conventions while SMS obeys *those*").

One reason we selected this relatively trivial example for analysis is that it brings to the surface some problematic assumptions about new media such as SMS for writing teachers. Because these technologies have been more widely adopted by other demographic sectors of the population (high school and college students as well as salespeople and executives), writing teachers frequently have difficulty understanding how genres are developed and evolve within them. One does not have to look far to see the old guard lamenting the rise of SMS and the related technology of instant messaging, characterizing the medium as the home of sloppy thinking skills, poor sentence construction, and misspellings. But on closer examination, we begin to notice that the conventions of the medium emerge relatively organically from the types of communication afforded by the technology and the contexts in which it is used in communication.

So, for example, Emily's e-mail was constructed in ways appropriate to that medium, including the social contexts in which it was used (an informal and quick message to a teammate). The content of the e-mail was relatively informal and designed to be quickly skimmed and responded to. The cell phone text message, on the other hand, was much more condensed, with intentional abbreviations and the use of single letters ("u" for "you") characteristic of the medium. These features are not due to sloppiness or poor writing skills—they are designed to accommodate a communication to the needs and affordances of texting, which supports very brief messages, a constraint arising from both the small size of cell phone displays and the difficulty of typing text into most cell phones (see Sun, 2004, on how users understand and adapt texting to their own, local needs).

That is not to say that SMS and IM communications are never sloppy or ineffective; but in many cases, automatically labeling them as such is a misguided critique. Emily's decision to use SMS is, for this context and desired change, probably the most appropriate.

EXAMPLE 2: CREATING SUPPORT FOR AN ON-CAMPUS CONFERENCE

In this second example, we'll take up a slightly more complex situation. The situation, which involves creating support materials for an academic conference, requires a host of communication media, of which wireless media emerges contextually as one important component.

Hannah, Steve, Amir, and Caitlin are gathered around a meeting table in the lab, hunched over memos, handwritten notes, booklets, printouts of web pages, several laptops, and other materials. Hannah sighs. "Where do we even start? How do we distribute all this information?"

"Let's back up a little," Amir says. "We're trying to reinvent the wheel. This isn't the first conference anyone has ever run. Why don't we start by looking at what other people have done for conferences like this?"

Steve pulls a laptop closer and brings a browser window to the front. He selects the organization's website from the bookmark window. "Look," he says. "They have archives of conferences going back to the early 1990s." He clicks on a link for the 1995 conference. The page opens to show a clunky-looking graphic, a brief paragraph of text discussing the theme of the conference, and a simple table of events, times, and locations. "Not much here."

"Try something more recent," Caitlin says. "Go to last year's conference."

Steve backs up and opens the page for last year's conference. At least they've improved the graphics. There's a lot more information as well.

"That's better," Hannah says. The page has links to the conference call for proposals, local information on hotels, a full conference program with links to abstracts, and more. They begin compiling a list of information, agreeing that nearly all of the categories on last year's site would be appropriate for this year's conference. As they sort through the list of items and take notes about where they can locate information relevant to the upcoming conference, they start talking about ways they might extend this year's offerings. "What about a live video feed of the main events?" Amir suggests. "Maybe," Steve says. "But who would view the video feed? And do we have bandwidth?" Caitlin asks, "Does it need to be live?" Amir agrees, saying, "It'll be easier to manage if we shoot video and then make it available after the conference." They discuss the possibility of an interactive video/audio feed that would allow people to participate, and then they decide they don't have a large enough staff to manage the complexities of it. Besides, they determine, the main users of the video/audio feed will probably be people using the information for reference purposes, perhaps showing selected sessions to classes.

They begin looking at the conference program, which they know people will use to decide what sessions to attend. The web pages for the program material run to several hundred, ranging from abstracts and speaker bios and pictures to maps, and more. "Do people really carry around this much paper while they're going to sessions?" Amir asks. "I guess so," Caitlin says. "But it seems like a pain. And how do we handle changes to the printed program? We have to get it to Printing Services at least a month before the conference, but things are going to change between then and the actual conference." "Can people just access the web site?" Amir responds. "We have wireless in the building, so if they have a laptop they can get updated information. Actually, we could give them all sorts of useful stuff on the site—like these maps of the building and the reviews of local restaurants. In fact, we could even do things like have online surveys about sessions to give speakers feedback. People could fill them out right after a talk and submit them online."

"What about PDAs? And cell phones?" Steve says. "That'd be cool. This is a conference for teaching with computers. We should push the envelope a little." "That's not really pushing the envelope," Hannah says. "Cell phones and PDAs have been around for awhile." "Yeah," Steve says, "But we're talking about teachers. They're just starting to figure out how to use their cell phones and PDAs." They all laugh, but they also know that there's more than a grain of truth in the joke.

"I haven't designed a web site for a cell phone before," Steve adds. "Have any of you?" They all shake their heads. "Huh. Is it hard to do?" "Can't be that hard," Caitlin says. "The display is small, so the pages will need to have less information. We'll have to think about what's important enough to include on the PDA and cell phone versions. And I'm not sure they use regular HTML. There must be some information on the Web about how to do it." She opens her own laptop and googles "cell phone web page design" and gets a bunch of pages on web site hosting. She then tries "designing web pages cell phone" and has more luck. The first page on the list has an overview of design issues for PDAs. She skims through it, summarizing key bits for the rest of the team. "Use basic HTML. Think small. No frames. Small graphics. Here's a link at the bottom for 'Wireless Markup Language'." She clicks it. "It looks like we can use WML to offer some additional things like text messaging. That would let people message each other during the conference if they wanted to meet up for dinner or something."

They spend fifteen minutes talking about other possibilities, following links, and bookmarking pages. Caitlin offers to spend some additional time doing research and figuring out useful ways to integrate WML into the conference before their next meeting. She also plans to follow up on some citations she noticed on the Web to "best practices" articles written by experienced WML designers. Then the team moves on to dividing up other work issues, such as who will gather information about local restaurants and who will code the draft program they currently have in Word format.

APPLYING THE FRAMEWORK

The example above takes a more explicit "designer" perspective on communication than the earlier library example. The C3T framework for the design of all of the conference materials would be relatively complicated, but if we focus primarily on the emerging decision to integrate WML into the conference materials, we can see ways that the framework captures (and could drive) decisions about design processes (Table 1.2). Note that in this case, we've omitted some aspects of the framework that deal with in-

TABLE 1.2. C3T CATEGORIES AND FEATURES FOR EXAMPLE 2

C3T CATEGORIES	FACTORS
CONTEXT	*Immediate*: Participants at an academic conference, both prior to arriving and also onsite, moving from session to session and in the local community. Wireless access via laptops, PDAs, and cell phones (but varies for attendees). *Broader*: Academics in teaching and research discourse communities.
CHANGE	Help users identify sessions to attend, correspond with other conference attendees, locate interesting local restaurants, etc. Notify users about changes to conference events.
CONTENT	Some existing material (conference program, abstracts, bios as gathered by conference program chair). Some material needs developing (restaurant reviews, ongoing program updates, etc.).
TOOLS	Basic web design and graphic design tools (both WYSIWYG and text editors). Resources with tips and references related to designing for PDAs and cell phones.

team communication. In some cases, it might be useful to consider such aspects in terms of team processes.

Context: The target context for this design project is fairly complex, dynamic, and varied. Because the students can't assume uniform access to any wireless technologies, they realize they'll need to design for multiple platforms. The context includes users who are highly motivated, but the users will need relatively streamlined information, both because of constraints in the form factor of PDAs and cell phones and because they will frequently be accessing information on the fly to locate rooms and activities. In the students' discussions, they seize on wireless mobile devices in part because of their "cool factor." For the particular users and contexts they're considering, such technologies make sense, although designers would also want to consider areas or needs that are better addressed (or additionally addressed) by print materials such as small reference booklets or signage.

Students will need to return to thinking about their users' context many times over the course of the project, as they gather additional information about specific changes, content they have available to them, and tools they'll be using (and that their users will be using).

Change: Users of this material should, ideally, experience change of at least two types: gaining information about conference events to decide which panels to attend, at which restaurants to meet, and so forth, as well as physical change in terms of moving physically to those locations. An additional level of change that the students are still exploring involves providing communication channels among conference attendees for things such as evaluating the quality of sessions and discussing issues raised during the conference (before, during, and after the conference).

In addition, at some point the student team would need to undertake some research on users of the materials they're preparing. Although they joke at one point about teachers being behind the curve on technology adoption, the situation might be worse than they think. For example, are the conference participants already comfortable using SMS? Will the cost of SMS (which is still, in the United States, relatively expensive compared to "free" Web access) discourage people from using the WML–enabled site they're discussing? Will they need to provide training materials?

Content: One primary feature of the content aspect involves identifying both existing and planned materials that will support targeted changes in context. For the WML aspects of the conference materials, students will be able to use some materials already gathered (or being gathered) by the conference program chair and assistant chair, such as abstracts, biographies, and contact information for attendees and speakers. Given the differences between larger display formats such as computer screen or print and PDA or cell phone, students will need to edit and chunk existing content, sometimes heavily, to match the specific traits of the communication technology. They will also need to learn how to code content for WML if they decide to use specific features of WML (such as SMS) that are not part of traditional HTML.

As they begin designing WML content, they also will need to research and design some information specifically for that technology. They will need to decide, for example, whether or not to include pictures of local restaurants (to support visually identifying buildings and locations for people unfamiliar with the local community), which would likely require taking digital pictures of restaurants, then editing them in various sizes to support print, Web, and WML. Similar content repurposing and creation will need to be undertaken for most aspects of the project.

Finally, students will need to revisit issues of context and change as they develop material, testing their evolving materials both heuristically (based on best practices concepts) and with actual users (to gather general information about user needs and perhaps to actually test specific pages and features with real users).

Tools: Although the students are relatively experienced designers of print documents and web sites and also have experience with using technologies such as SMS and WML, they have no experience designing for wireless devices apart from those used to access traditional Web content on standard computer screens. So one initial phase of their work will involve learning tools (both software and syntax) for WML. Their initial investigations suggest that the learning curve will not be abnormally steep, but they may discover that such learning is not worth the return on investment (particularly when they return recursively to thinking about their users in context, who may not be inclined to actually access the WML support materials). In addition, advanced features for some (but not all) wireless carriers require expensive developers' tools.

CONCLUSION

In this chapter, we've provided a basic framework for situating communication design. As we indicated earlier, our goal here was not to provide a framework only for mobile wireless technologies. The rapid pace of technological change would undoubtedly limit the useful shelf life of such advice. Instead, we've attempted to outline a framework that can accommodate a variety of constantly changing technologies based on investigating key features of context, change, content, and tools. This extensible and open framework itself is built from relatively time-honored and tested approaches to understanding and teaching communication design.

As we've said already, the framework assumes (but does not require) a more complex approach to writing situations than sometimes covered in writing courses. By organizing coursework around realistic projects, students are given the opportunity to work with a wide range of tools, contexts, and content in flexible and varied ways. This complexity provides them with a richer sense of communication as an active, ongoing process involving multiple people in specific contexts. Project-based courses are, in this way, a bridge between classroom learning and real-world activity, frequently oscillating between instruction or mini-lecture and application and critique. The wide net cast by the framework can also be used to help students understand their own knowledge and skills as communicators—most of them have been communicating effectively for years in various contexts. The classroom then becomes a place where they learn a framework that can help them learn new things that they can add to their repertoire. More importantly, the framework can provide them not simply with sets of skills suited to specific, static situations, but a method for adapting those skills to new situations.

ENDNOTES

1. The images in this chapter were taken with cell phone cameras as we traveled to conferences, leading to relatively poor image quality compared to dedicated cameras. As authors, we struggled to decide whether or not the images were of high enough quality for publication. This decision process provides an example of the rhetorical decisions writers must make in working across rhetorical and technological contexts. Whether or not we made the right choice in including them here is still an open question for us.
2. See note 1.

REFERENCES

Baron, Dennis. (1999). From pencils to pixels: The stages of literacy technologies. In Gail E. Hawisher & Cynthia L. Selfe (Eds.), *Passions, pedagogies, and 21st century technologies* (pp. 15-33). Logan: Utah State University Press.

Birkerts, Sven. (1994). *The Gutenberg elegies: The fate of reading in an electronic age.* Boston: Faber and Faber.

Eisenstein, Elizabeth. (1979). *The printing press as an agent of change.* New York: Cambridge University Press.

Futures: Meeting the challenge. (2005). *Qualifications and curriculum authority.* London: QCA. Retrieved November 27, 2005, from http://lists.ncte.org/t/605 427/686481/2836/0/ http://www.qca.org.uk/14303.html

Hawisher, Gail E. (1988). Research update: Writing and word processing. *Computers and Composition, 5,* 7-27.

Heba, Gary. (1997). HyperRhetoric: Multimedia, literacy, and the future of composition. *Computers and Composition, 14,* 19-44.

Kemp, Fred. (1992). Who programmed this? Examining the instructional attitudes of writing-support software. *Computers and Composition, 10,* 9-24.

McCarter, Mickey. (2003). Instant messaging finds military market. *Military Information Technology, 7.* Retrieved March 22, 2005, from http://www.military-information-technology.com/article.cfm?DocID = 110

Reid, Rebecca. (2004). IM emerges from the shadows. *PC World.* Retrieved March 22, 2005, from http://www.pcworld.com/news/article/0,aid,114322,00.asp

Sanders, Barry. (1994). *A is for ox: The collapse of literacy and the rise of violence in an electronic age.* New York: Vintage Books.

Selfe, Cynthia L. (1989). Redefining literacy: The multilayered grammars of computers. In Gail E. Hawisher & Cynthia L. Selfe (Eds.), *Critical perspectives on computers and composition instruction* (pp. 3-15). New York: Teachers College Press.

Sun, Huatong. (2004). *Expanding the scope of localization: A cultural usability perspective on mobile text messaging use in American and Chinese contexts.* Unpublished dissertation, Rensselaer Polytechnic University.

Yancey, Kathleen Blake. (2003). The pleasures of digital discussions: Lessons, challenges, recommendations, and reflections. In Pamela Takayoshi & Brian Huot (Eds.), *Teaching writing with computers: An introduction* (pp. 105-117). Boston: Houghton Mifflin.

2

LEARNING UNPLUGGED

Teddi Fishman
Kathleen Blake Yancey

WHAT ROLE TECHNOLOGY?

Historically, it is rare for students to be introduced to technology by their teachers and schools. Rather, it is more common that already available technologies in commercial, entertainment, or media applications e-v-e-n-t-u-a-l-l-y make their way into instructional use after having already become familiar in other contexts. Sometimes, as in the case of television in the 1970s, that delay has been good news. By the time television eventually became common and affordable enough to make its way into a significant number of classrooms, we had already discovered that "turning on and tuning in" lacked the interactivity necessary to be pedagogically effective. In the case of television, then, the lag time between the commercial and academic applications of technology was a good thing: With relatively few TV sets in the classroom, we used them for less than we might otherwise have done had we immediately embraced (and therefore invested heavily in) TV as a pedagogical device. In contrast, however, because it is highly interactive, ubiquitous, and persistent, wireless technology shows both more tenacity and more promise, ironically perhaps, as a classroom fixture and a potential pedagogical means.

Although wireless technology may be the latest instantiation of what's cool, it too has operated on the everywhere-else-first-and-and then-per-haps-in-school principle. This principle becomes acutely obvious when we recall that years before most classrooms went wireless (and most in fact aren't wireless yet), students created their *own* wireless zones with their cell phones (as anyone who has ever tried prohibiting the ringing of cell phones in the classroom has learned) and pagers (who needs a network hub?), bringing into our *inside spaces* their own *connections to the world out-side* and their own *information and social networks*. Those connections— the ones we don't invite, control, or even welcome—are typically perceived by faculty at best as a nuisance, at worst as a breach of classroom decorum and an affront to education itself. In fact, the immediate reaction from many if not most faculty is located precisely in the dichotomy between the world of higher education and the "real world" outside: Wireless—despite its potential for making connections, accessing information, and supplying "just-in-time knowledge"—is *out*.

To be fair, many faculty are as least as frightened as they are affront-ed, and not without reason. A genuine concern is located in the question, "If wireless is part of the new pedagogical landscape and can supply *all* of the answers, why will students continue to come to class?" In fact, why should *I*? In the world of wireless, what's the exigency for face-to-face ped-agogy? This question, of course, prompts the larger question: "What's the exigency for *my* pedagogy in an age where virtually anything that is known can be remediated and accessed anywhere, anytime?"

WIRELESS AS A VEHICLE TO EMPHASIZE PROCESSES OF KNOWING

To the extent that there is an answer to the question of pedagogical exi-gency, it lies in the potential for wireless to allow educators to focus on high-er level processes of engagement with, analysis of, and manipulation of data rather than simply recollecting and regurgitating information (which is related, of course, to whether wireless in the classroom is good news or bad news in a given instructional setting). How it's perceived, as a risk or a ben-efit, and what effects it exerts, beneficent or obstructive, depend both upon pedagogical approach and upon the way one views education itself.

Put simply, it's about learning, about how we construct it and how we "deliver" it. If the course goals and methods of evaluation center on rote knowledge, memorization, and/or the acquisition of "content knowledge," the risk posed by unauthorized access to information can be significant. Such a pedagogy tends to be located not in students, but rather in the expertise of the teacher. When expertise, however, is addressed as a resource and the teacher as a guide to locating, interpreting, evaluating, and using information, the focus of classroom activity necessarily shifts. When the availability of virtually unlimited information is a given, but the quality, applicability, veracity, and authenticity must be investigated, pedagogy changes from "what do you know?" to "how do we know?" or "how can we know?" or "what can we do with what we know?" Particularly when students are expected to generate information as well as consume it—processing, manipulating, and creating—access to multiple resources can expand, deepen, and complicate both what is known and how what is known *comes to be known.*

Clearly, when wireless becomes integral to the classroom environment, methods of teaching must change, and so too must the methodology of assessment. An exam that asks students merely to list the elements of the periodic chart, for example, will be more vulnerable to compromise by access to information afforded by wireless technology, just as it was vulnerable to students who would write the information in tiny letters on their wrist or the underside of a ball cap. In contrast, an assessment that asks students to *use* the information on the chart—by comparing the suitability for specific industrial uses of certain elements on the top of the chart to others at the bottom, for example—might find in wireless a distinct advantage. The likelihood of academic dishonesty is lower, but perhaps more importantly, the usefulness of the task as preparation for the work they will eventually do with this information is significantly greater. In contrast to tests of recall and predetermined application, *this* model of assessment allows students to showcase the very skills and knowledge creation that will be useful inside *and* outside of school. And in the case of skills-based or performance-based evaluations, what also can be assessed is the ability of the person being tested to identify, locate, and utilize the appropriate information to complete the task, competencies that are part of Information and Communication Technologies (ICT), the primary literacy of the 21st century, and a new subject of composition.

Coming to Class

The prospect of wireless-enhanced pedagogy that assumes unlimited access to information and focuses almost exclusively on means of analysis, critical and evaluative approaches, and methods of data processing (including locating, evaluating, and using in multiple ways) rather than on content information *should encourage* a re-examination of why (or whether) students should come to class. As access to people as well as to information becomes less and less related to physical space and increasingly about networks, connection speed, and compatibility, it may be that it no longer makes sense to place much importance on being in a particular place at a particular time. Traditional attendance policies, for instance, may morph into participation policies wherein "being there" means being there for a purpose—to help, to contribute, to share—in short, to *do* rather than simply *be*.

Currently, of course, instructors take attendance (if they do) to encourage class attendance, and students often comply, often out of fear that they might miss *the* one class day standing between them and a "C." Some students, faculty have found, will believe class superfluous if class notes and slides are posted online. After all, as these students proclaim through their absence, if the material is available online, why get out of bed and drag one's body to class at 8:00 a.m.? (No wonder faculty worry about the effect of wireless.) At the same time, anecdotal information suggests, at least in the case of the sciences, that this view of classroom-as-unnecessary, even in the case of a large lecture classes, is wrong (Julia Frugoli, personal communication, 2000). Students relying on this method of learning don't in fact perform well on the course assessments. In other words, although we might argue about how to define learning, whatever *it* is seems to require more than reviewing slides that are intended as a supplement to class. But perhaps the problem is just that, that the slides are supplements. What happens when the slides *are* the course, much as we see in the college-course-in-a-box at Barnes and Noble?[1]

Of course, even if the slide-review-qua-learning does "work"—does enable students to learn, to critique, and to create—the class itself, with a physical location, still serves a function (and this is true even in other models of education, as in the United Kingdom and Scandinavia): testing. On days when tests and assignment due dates are scheduled, all students are required to *appear*. In the world of wireless, is that all that will be left for f2f to provide? Even as testing specialists scramble to devise innovative ways of assuring that students cannot beat the system with surrogate test-takers, students find alternatives to attending testing sessions in courses where they are known only as a number. Given the nature of testing technologies and the complexities of password protections, how long will the classroom even serve this function?

WIRELESS AS TETHERED TECHNOLOGY

One time, the phrase "classroom practices" referred to activities conducted while teachers and learners inhabited the same physical space. Today, the phrase "classroom practices" increasingly serves as a synecdoche for the wide range of activities that happens not just within the confines of a particular room or during specified hours, but which spills over and out, into other times (out-of-class hours); other places (cyberspace); and if we are lucky, other lived spaces (students' "real lives").[2] In this version of learning, such seemingly transparent terms as discussion, collaboration, and group work are constantly redefined to accommodate the new forms that these activities take when they are computer-mediated activities. When the classroom itself reaches beyond its own boundaries, the effect is compounded.

Consider a specific case. On the Clemson University campus, the architecture program has been redesigned to be a "fluid" program. What this means is that students can take courses at one or more of four campuses—Clemson, S.C.; Charleston, S.C.; Genoa, Italy; and Barcelona, Spain. Critical to this new curricular offering is the interface provided by the virtual, and this, the architects say, is a challenge. Although the details of allowing students to work and learn in a number of geographical spaces have been worked out, the issues related to the virtual spaces that serve as common ground (and how serious a pun is this metaphor?) are still unresolved. Hardware and software compatibility issues, time zone differences, and linguistic barriers between and among humans and machines all contribute to the complexity of communicating in online environments, which (ironically) attract us specifically because they are free from physical constraints.

Put another way, wireless is not untethered; we know this empirically, from our own experiences, as well as from Verizon's "Can you hear me now?" advertising campaign. It is embedded in a physical context and even more so when the physical contexts are multiplied; it can serve as interface function; it can serve as its own site of learning. This means, most obviously, that in order to "work," wireless still requires certain physical proximities—to a signal source or a power source. It may also mean, however, that other kinds of proximities are also required—proximities of discourse, proximities of perspective, proximities of knowledge. It may be that as the power sources—and thus wireless itself—become more easily available, the latter become the issues deciding the ultimate success or failure of communication in a given context.

WIRELESS, CLASSROOMS,
AND MULTIPLE CONTEXTS

According to the Portraits of Composition Study (Yancey et al., 2005), discussion is a staple of writing classrooms. How does wireless change that?
Presumably, students are still discussing, but with whom? Is it possible that
our concern about wireless is located, at least in part, with the capacity for
wireless technology to subvert our authority in yet another way? Presently,
we determine whether classroom discussion happens with a partner, within small groups, or as is happening with increasing frequency, across multiple sections of the same course, and we decide who the partners are, who
is grouped with whom. Wireless changes this control as well, allowing students to expand their reach to anyone—peer students across the world,
younger students across town, outside experts (who *may* disagree). For the
first time, instructors must decide, how important it is (if it is important at
all?) that these discussions remain anchored in the classroom? And for
what purpose, limiting or extending the conversation?[3] If students' impulse
is to bring "others" into the writing classroom—via text messaging, e-mail,
or even cellular phone—is this an impulse we should (continue to) resist,
or should we find ways to turn these "intrusions" into contributions? It may
be that at least part of the threat of wireless communication stems from its
tendency to exclude the teacher (who is rarely a recipient of or participant
in extra-classroom communications during class time). If that is the case,
perhaps the solution lies in making those conversations *more* rather than
less salient in the face-to-face (f2f) classroom.

   Although being "wired" brings attention to the ways in which the
boundaries that demarcate instructional space are blurry, inconsistent, and
imprecise, the residual assumptions and expectations accompanying the
classrooms with which most of us are familiar persist (Zoetewey, 2004).
Perhaps this is because the changes in computer technologies, particularly
in the classroom, have occurred incrementally. Word-processing stations
gave way to standalone computers, which were supplanted by intranets
and networked classrooms. When technology was limited to performing
simple functions (typing and formatting), the information it could provide
was limited to what came with the machine, and except in some few cases
(Takayoshi & Blair, 1997), we didn't attend to differences in processes and
products. "Wired" classrooms, and eventually wireless zones, provided a
corollary increase in the range and scope of media access that the classroom could furnish, as it made the classroom more complex in its degrees

of mediation. The usefulness of digital technology increased exponentially as computers became linked with each other and could share documents and applications. The usefulness (and significance) of information technology increased again as computers in the classroom became linked to the outside world via wired connections. What we are witnessing now may tell us what further gains can be made as learners become part of a real-time information web, as we continue the transition from wired to wireless.

* * *

In thinking about how mobile communication technologies are altering the world as we know it, William Mitchell (1999) talks about overlapping spaces. His claim is that primary relationships—with family and close friends—will not be changed. What will be changed is what he calls our secondary relationships, with those we work with. Specifically, he says that for some time we will experiment with a mix of online and place-based community work.

For some of us, that time is now.

WIRELESS AND ARTIFACTS

According to the tenets of activity theory, human actions are mediated by artifacts that are instrumental in the pursuit of performative objectives. Because these artifacts become tools only through activity, context becomes a primary consideration in any examination of a particular technology. When viewed through this lens, wireless communication technologies can function on one (or more) of three levels: (a) authentic activities (which are conscious and motive-oriented) such as communicating in a foreign language; (b) actions, which are more oriented toward specific goals such as searching for a particular vocabulary word in online database; and (c) operations, which are usually not conscious and which relate to more routine behavior such as performing the keystrokes to enter search terms.

A second consideration with respect to activity theory is the extent to which tools and users affect each other mutually and continually. Clearly, even though it is a relatively new technology, wireless has evolved to suit the needs and purposes of its users. It is even clearer that our behavior has changed in response to the availability of cell phones, Global Positioning Systems (GPS), and integrated communication systems. What can be done—text messaging, remotely logging into a database, bypassing a centralized, institutional computer network in favor of an independent service provider—influences what is done, which in turn affects the next round of hardware and software development.

What Does All This Mean for the Classroom?

For the classroom on a commuter campus, particularly one where the student body is typically dispersed to home and job, it means that collaboration might best take place outside class and online. Current data—those of the Community College Survey of Student Engagement (CCSSE), for instance—show quite clearly that at the reporting institutions, some 400 community colleges, students have quite different experiences in class than out of class. For example, while students report substantial in-class collaboration and group work, only 21% of those students report analogous out-of-class activity (CCSSE Survey Results, 2005, p. 15). Such results are no surprise if we are asking such students (who often have jobs and families) to gather f2f for outside-class activity. With wireless, they could gather virtually. And pragmatically, what this means is that collaborative activities assigned for outside class that were difficult to complete because of material conditions can now become (an equal) part of the curriculum. And what that means is no less significant: The curriculum changes.

For the classroom on a residential campus, it means much the same. Students can exercise more choice: meeting physically, meeting electronically. Students who like to stay up late but who want to work in their separate dorms can work as a group online. Coffee houses and botanical gardens become sites of learning, through the virtual interlaced and interfaced with/in the classroom setting.

For the researcher, wireless raises questions about which tasks lend themselves better to f2f communication; which to electronic; and which to both. It may be, for example, that we are all so different that no general principles will hold. It may also be that certain tasks, for example, planning a project, lend themselves well to f2f interaction, whereas other tasks, such as updating group work, are better completed online precisely because a record is created. It's also so that if we choose to include wireless as one of our major modes of delivery, we may need to modify our curriculum to ensure that students have a chance to learn *how to learn* in this environment, and as we learn about this site of learning, we may better understand the f2f classroom-learning environment as well. How will students seek to transfer what they learn in one site to the next? What do we want them to transfer? Questions about transfer, as David Smit (2005) argues, are critical to the new composition of the 21st century, and they are likely to play out—and mean—differently in these overlapping sites.

Texture and Substance

How to learn, increasingly, involves context. In the day of the classroom box (Yancey, 2009), the text was what mattered, that and the instructor's interpretation. Context, in fact, was expertise. Right now, in both popular and high culture, context belongs to us all—which of course is the claim of wireless.

Consider, for instance, the new design of the Museum of Modern Art. Described by one critic as a hypertext, it organizes some art by historical context, other by juxtaposition, other by a need for space. Even the context for the art, the building, nearly disappears, as John Updike (2004) explains:

> Nothing in the new building is obtrusive, nothing is cheap. It feels breathless with unspared expense. It has the enchantment of a bank after hours, of a honeycomb emptied of honey and flooded with a soft glow. My guide, William J. Maloney, the genial project director, quoted the architect as saying to the museum trustees something like this: "Raise a lot of money for me, I'll give you good architecture. Raise even more money, I'll make the architecture disappear." And disappear, in a way, it has. (¶ 2)

The art is there, of course. But the Museum of Modern Art is postmodern in its celebration of absence; very little direction or explanation is provided, and the art is, one finds, not in but out of order. In other words, you're as likely to have a Picasso next to a Jasper Johns as to a Matisse. The result is a museum, which like wireless, offers multiple experiences under one roof: Again, John Updike (2004):

> With the expansion of 1964, which added the great Picasso-Matisse room, some choices for ambulation were offered; but it was still, on the second floor, a single experience. Now four floors, plus soundproof galleries for video and media, beckon from all sides. (¶ 14)

What this *arrangement* means is that the museum-goer is an *inventor*, too, filling the absence with as much and many contexts as possible.

And it's not just high art: We see a similar effect even on TV, the most popular of pop culture. According to Stephen Johnson (2005), TV is contributing to what he calls a smart culture.

> Over the last half-century, programming on TV has increased the demands it places on precisely these mental faculties. This growing complexity involves three primary elements: multiple threading, flashing arrows and social networks. (¶ 7)

Of particular interest to us is the point he makes about context and the increased demands placed on viewers. Through the 1980s, Johnson argues, TV (even "good" TV like "The Mary Tyler Moore Show") presented a straightforward narrative complete with the preferred context. Today, he points out, TV presents narratives that extend beyond one show, that include multiple subplots, and that require the viewer to complete the (contextual) picture. Consider, he says, "The West Wing."

> A contemporary drama like "The West Wing," on the other hand, constantly embeds mysteries into the present-tense events: you see characters performing actions or discussing events about which crucial information has been deliberately withheld. Anyone who has watched more than a handful of "The West Wing" episodes closely will know the feeling: scene after scene refers to some clearly crucial but unexplained piece of information, and after the sixth reference, you'll find yourself wishing you could rewind the tape to figure out what they're talking about, assuming you've missed something. And then you realize that you're supposed to be confused. The open question posed by these sequences is not "How will this turn out in the end?" The question is "What's happening right now?" (Johnson, 2005, ¶ 19)

Johnson (2005) offers two terms that help us differentiate what TV—and education?—offers: texture and substance.

> Popular entertainment that addresses technical issues—whether they are the intricacies of passing legislation, or of performing a heart bypass, or of operating a particle accelerator—conventionally switches between two modes of information in dialogue: texture and substance. Texture is all the arcane verbiage provided to convince the viewer that they're watching Actual Doctors at Work; substance is the material planted amid the background texture that the viewer needs make sense of the plot. (¶ 20)

If there is a question posed by wireless, it's surely, "What's happening right now?" And if context is important in the world of wireless, what counts as texture and what as substance?

Can You Hear Me Now?

A second, related conceptual framework that may be helpful is actor-network theory, which similarly acknowledges the ways in which technologies reflect the histories and experiences of users as well as previous generations of technology. As the name implies, however, this approach places more emphasis on networks of human and nonhuman stakeholders.

Although the term "wireless" implies a lack of physical tethers, it does not follow (and in fact is clearly not the case) that wireless technologies are not subject to physical boundaries. As Verizon's "Can you hear me now?" advertising campaign humorously illustrates, there are abrupt albeit invisible lines of demarcation that separate "in" from "out" as surely as concrete walls. Perhaps the most significant change is that, in (or out of) wireless environments, it is not nearly as obvious on which side of the wall one is. Wireless technology also raises expectations of freedom that are not necessarily met. In fact, the possibility of connection (to a network, to global telecommunications) is in many ways at least as restrictive as those imposed by physical connections.

WIRELESS AS DYSTOPIAN

For faculty who build community in classrooms, wireless seems a threat rather than a promise. It's not easy to teach virtually. Richard Courage (2006), for example, shows how easy it is to misinterpret in the online class, and how the dynamics there are very different than in f2f classes—and for many students, not better.

And that assumes that wireless doesn't simply shut down whatever it is that we think teaching and learning are. I think of a student we know, a double major in engineering and math at Virginia Tech, whose 3.6 G.P.A. suggests that he has mastered the art of being a successful college student. Currently, he is taking a course in alternate energy systems. He was, he says, hoping for a "hippie-style" professor who would help students learn to advocate for alternative systems. The professor teaching the course, it turns out, is a corporate type more interested in helping them learn the art of cost analysis for large utility companies. And that's not the only disappointment. The class has met exactly once, early in the term, when the faculty member reviewed the syllabus. Since then, every class activity, other than due dates for tests, has been asynchronous; when asked to clarify concepts or procedures, the professor refers students to the e-published class notes; and there is no book for the course because the professor has not finished writing the one he wants the students to use. In sum, whatever this class is about, it's not learning—and we should note, this kind of class is only made possible by digital technology. Put another way, even distance education by correspondence assumed some correspondence, some exchange between teacher and student.

But a change in medium can be a change in everything: That's not to say that bad teaching is caused by wireless or digital technology generally; it is to say that this kind of bad teaching is. But there's a less dystopian view. Perhaps there's a vision of teaching and learning where wireless does enhance learning, and perhaps too the way we might want to work today is already in the making. I think about classes where listserv assignments are as common as discussions. I think of a colleague in chemistry who begins f2f class with students using handhelds to beam ("click") answers to questions to a central location so that he can deliver what students don't know. I think about the meeting I just went to, where this same colleague recorded our notes on a tablet PC and beamed those to a Smartboard and saved them as a .jpg and sent them on to us virtually, too. These different means of communication are not in competition with each other, but in cooperation.

How might we plan teaching, learning, and curriculum to take advantage of such cooperation?

UTOPIAN IMPULSES

Although it may be true that end-users don't conceptualize wireless as bound by physical spaces, that is not the case for system administrators and providers, as evidenced by battles that are being waged in the courts over who has the right to offer wireless services in specific areas. In Philadelphia, PA, even though high-speed internet connections are not available to most residents, the city was restrained from making free wireless available in order to protect the interests of wireless service providers who claim that unless their interests are protected, they will not invest in the infrastructure required for continuing expansion (Lessig, 2005, p. 80). Even where wireless is more widely available, differences in bandwidth availability (which translates into speed) stratify wireless functionality. Wireless, in other words, is not a valid response to a utopian impulse.

THE RELEVANCE OF WIRELESS

In the 1960s a cry for relevance went out, in 2002 NCTE published *The Relevance of English*, and in the last 40 years, if judged by the number of English majors of both undergraduate and graduate varieties, we have become not more relevant, but less so. To counter this, some English departments are retitling themselves, from Departments of English to Foreign Languages and Telecommunications (Alabama A&M); others (like George Mason University) highlight three themes: "Fall for the Book," "Technology in English," and "Text and Community." Still others, such as Duke University's iPod program, use student affection for hardware to make new links. But many of us have not incorporated digital technology into our outcomes for composition, and even many of us with wireless access are uncertain as to how to proceed. In the interim, as *Newsweek* points out, many *high school* students already learn in an environment where "Everything is up for grabs: curriculum, size, even the idea of school itself. With new technology that puts the world at their keyboards, students can learn without a classroom or formal teacher" (Kantrowitz, 2005, ¶ 1). Of course, Peter Elbow made this argument nearly 30 years ago, but in a quite different way.

What that way means *now* is what we need to decide. And given the speed of both wireless and technological change, we'd best be quick about it.

ENDNOTES

1. In the fall of 2003, the University of Florida began delivering first-year composition by replacing small classes, each with an instructor, with a large lecture linked to a set of PowerPoint slides. And as Darin Payne (2005) argues in a recent *College English* article, Blackboard as a content management system is a version of a course in a box.

2. Those familiar with current work in composition spaces—e.g., that of Nedra Reynolds (2004)—will find resonances of Soja's thirdspace here. Although third-space is an intriguing concept, it fails to account for the built environment, and as more than one architect has noted, architecture is rhetorical: Through its definition of space, it fosters some relationships and discourages others. As important, perhaps, what we are witnessing now—and participating in—relative to education is a shift from the rationalism of modernism to the multiple sensibility of postmodernism, as noted by architectural critic Herbert Muschamp (2003): "Rationalism made a place for transcendence in the modern city: it is a drilling platform for expectation. Properly positioned, the apparatus of reason lets contemporaneity bubble up from the depths of urban nature" (np). Wireless is by analog contemporaneity.
3. It's fair to say that these questions are not unique to wireless situations. In discussing the various pedagogical uses of listservs, for instance, Yancey (2002) specifically points to their value to bring in outside experts. Still, the assumption thus far is that the faculty members will make the assignment, will approve the expert, and will otherwise control the situation. Not so with wireless, at least potentially.

ACKNOWLEDGEMENTS

The authors would like to thank Karen Kaiser Lee, a graduate student at Florida State University, for her editorial assistance.

REFERENCES

Community College Survey of Student Engagement (CCSSE). (2005). Engaging students, challenging the odds. Retrieved April 17, 2006, from http://www.ccsse.org/publications/CCSSE_reportfinal2005.pdf

Courage, Richard. (2006). Asynchronicity: Delivering composition and literature in the cyberclassroom. In Kathleen Blake Yancey (Ed.), *Delivering college composition: The fifth canon* (pp. 168-183). Portsmouth, NH: Heinemann/BoyntonCook.

Elbow, Peter. (1973). *Writing without teachers*. Oxford: Oxford University Press.

Johnson, Steven. (2005, April 24). Watching TV makes you smarter. *New York Times Magazine*. Retrieved April 17, 2006, from http://www.nytimes.com/2005/04/24/magazine/24TV.html?ei = 5088&en = e08bc749e56cbb59&ex = 1271995200&pagewanted = print&position

Kantrowitz, Barbara. (2005). The 100 best high schools in America. *Newsweek*. Retrieved April 17, 2006, from http://www.msnbc.msn.com/id/7761678/site/newsweek/

Lessig, Lawrence. (2005). Why your broadband sucks. *Wired, 13,* 80.

McBride, Neil. (2000). Using actor-network theory to predict the organizational success of a communications network. Retrieved March 3, 2006, from http://www.cse.dmu.ac.uk/~nkm/WTCPAP.html

Mitchell, William J. (1999). *E-topia.* Cambridge, MA: MIT Press.

Muschamp, Herbert (2003, June 16). Blond ambition on red brick. *The New York Times.* Retrieved April 14, 2006, from http://massengale.typepad.com/venustas/2005/10/muschamp_on_new.html

Payne, Darin. (2005). English studies in Levittown: Rhetorics of space and technology in course-management software. *College English, 67,* 383-408.

Reynolds, Nedra. (2004). *Geographies of writing: Inhabiting places and encountering difference.* Carbondale: Southern Illinois University Press.

Smit, David. (2005). *The end of composition studies.* Carbondale: Southern Illinois University Press.

Takayoshi, Pamela, & Blair, Kristine. (1997). Reflections on reading and evaluating electronic portfolios. In Kathleen Blake Yancey & Irwin Weiser (Eds.), *Situating portfolios: Four perspectives* (pp. 357-369). Logan: Utah State University Press.

Updike, John. (2004). Invisible cathedral: A walk through the new Modern. *New Yorker.* Retrieved March 3, 2006, from http://www.newyorker.com/critics/atlarge/?041115crat_atlarge

Yancey, Kathleen Blake. (2002). The pleasures of digital discussions: Lessons, challenges, recommendations, and reflections. In Brian Huot & Pamela Takayoshi (Eds.), *Teaching writing with computers: An introduction* (pp. 105-117). New York: Houghton Mifflin.

Yancey, Kathleen Blake. 2009). Spaces of composition. In Douglas Reichert Powell & John Paul Tassoni (Eds.), *Composing other spaces* (pp. 203-216). Cresskill, NJ: Hampton Press.

Yancey, Kathleen Blake, Fishman, Teddi, Gresham, Morgan, Neal, Michael, & Taylor, Summer. (2005, March 18). *Portraits of composition: How writing gets taught in the early twenty-first century.* Paper presented at the College Composition and Communication Conference, San Francisco, CA.

Zoetewey, Meredith. (2004). Disrupting the computer lab(oratory): Names, metaphors, and the wireless writing classroom. *Kairos, 9(1).* Retrieved January 4, 2006, from http://english.ttu.edu/kairos/9.1/binder2.html?coverweb/zoetewey/index.html.

PART II

Examining Teacher and
Student Subjectivities in
the Age of Wireless and
Mobile Technologies

3

"A WHOLE NEW BREED OF STUDENT OUT THERE"

WIRELESS TECHNOLOGY ADS AND TEACHER IDENTITY

Karla Saari Kitalong

AT&T's "You Will . . ." ad campaign, which aired in 1993 and 1994, fore-shadowed the nearly ubiquitous wireless communication that we enjoy today. AT&T predicted (correctly, I might argue) that wireless technology would revolutionize business, medicine, family life, and education. Each of the 30-second spots in the ad campaign asked a trio of "Have you ever . . ." questions such as the following:

Have you ever . . .

- Had a classmate who was thousands of miles away?
- Learned special things from faraway places?
- Attended a meeting in your bare feet?
- Sent a fax from the beach?
- Carried your medical history in your wallet?
- Tucked your baby in from a phone booth?

Regardless of which three possibilities it posed, each commercial answered its questions in exactly the same way: "*YOU WILL. And the company that will bring it to you: AT&T*" ("Banner ads," 2004).

Some of the possibilities AT&T envisioned have come to pass, although none (yet) rely exclusively upon wireless technologies. Today, for example, online learning opportunities are based both in traditional schools, colleges, and universities and in for-profit institutions such as the University of Phoenix that exist almost entirely in cyberspace. Meanwhile, content available on the Internet free of charge (once one has access to the necessary technology) facilitates self-directed learning on nearly any topic imaginable. Telecommuting is a viable practice for many employees, who sometimes *do* conduct web cam meetings from their home offices as AT&T predicted they would. On the other hand, live streaming video by which to tuck in a baby from a distance has not become a feature of phone booths. In fact, cell phones have rendered phone booths all but obsolete. Additionally, despite increased concern for patient privacy, emerging practices such as teleradiology allow certain medical tests to be read remotely (Friedman, 2005, p. 16). Finally, in today's 24/7 world, the ability to send a fax from the beach conjures up not only the convenience of ubiquitous technology but also the unwelcome intrusion of ubiquitous work into family life and leisure time (Kitalong, 2000). AT&T's futuristic vision, now called Mobile Life (M-Life for short), has arrived, at least to some extent, albeit accompanied by a host of advantages and disadvantages.

In this chapter, I examine how teachers are portrayed in advertising and editorial representations of the wireless world AT&T predicted a decade ago. To identify suitable ads, I examined the 2003, 2004, and 2005 issues of three educational technology periodicals (*Syllabus, Campus Technology,* and *Edutopia*) that are offered free of charge to teachers and other school personnel. Specifically, I looked for ads that depicted or promoted wireless technology in educational contexts, and that featured teachers as the *represented participants* (the depicted subjects). However, it quickly became apparent that few teachers appear in ads showcasing wireless technology, although teachers can sometimes be inferred to be the *interactive participants* (the targeted audience members) (Kress & Van Leeuwen, 1996, p. 46).

Before I had analyzed a single image, then, I was forced to conclude that teachers are given short shrift in such media representations. I argue that the media representations I discuss in this chapter constrain what I call teachers' *identity potential*. I define identity potential as the repertoire of available beliefs and behaviors from which an individual can draw, based on factors such as race, gender, ethnicity, subcultural affiliation, educational and profession attainments.[1] Of course, teachers cannot change the media representations that help to construct their identity potential, but they can reinterpret those representations. I conclude this chapter by demonstrating how teachers can take ownership of their identities by adopting Stuart Selber's (2004) three-part computer literacy framework. Selber intends this framework to serve as a theoretical and practical model for teaching digital literacies to college students; however, I suggest that it is potentially also a toolkit teachers can employ to help broaden their own identity potential within the conceptual landscape of technological literacy.

The advertisement for Capella University's online master's degree in education (see Figure 3.1) sparked this line of inquiry for me, because it sends a message typical of the ads found in technology periodicals such as *Syllabus, Campus Technology,* and *Edutopia*. Although it doesn't directly attempt to sell wireless technology, it does represent an educational realm in which wireless technologies are the norm.

This ad appeared in several issues of *Syllabus* in late 2003 and

Figure 3.1. Capella University Ad

early 2004. The interactive participants for the ad—its audience members—are teachers who believe their technology skills to be lacking. The ad's represented participant is a little girl wearing beaded bracelets and a pink sweater. Her small hands rest on a wood-laminate school desk on which are arranged traditional writing implements (a colorful pencil, a small notebook). She is engaged in a classic elementary school activity—note passing—but with a difference. She writes her note not with a pencil on a scrap of paper but with a stylus on a personal digital assistant (PDA). The note reads as follows:

I like you. Do you like me?
Yes ❏   No ❏

The ad's tag line, "Because there's a whole new breed of student out there," appears in a box that resembles a popup ad on a computer screen. Indeed, the pink-sweatered girl whose dimpled, bracelet-adorned hands hold the PDA can be seen as a "new breed of student." Certainly, she is no older than 10, which means that she was born after AT&T's "You Will . . . " campaign hit the airwaves. In other words, she has always lived in a world in which instant wireless communication could be envisioned, if not literally carried out. The ad reveals her point of view, not the teacher's, signifying that PDAs are the tools of her world, a world not limited by the classroom walls or confined to the teacher's lesson plan. She has abandoned her traditional writing tools. The "you" to whom she directs her message could be sitting on the other side of the room or on the other side of the world.

The student's identity potential is expansive: She is an experienced wireless network user who operates simultaneously both inside and outside the classroom with traditional and electronic communication tools. Although she occupies a seat in the classroom and is apparently doing her schoolwork, she is at the same time socializing with someone who may not be in the same room or even on the same continent. The ad gives the teacher much less leeway with respect to her identity potential. When my third-grade teacher, Mrs. O'Neill, caught students passing notes, she responded, predictably, by confiscating the offending missives. Sometimes she read them out loud to the entire class—or threatened to do so. The prospect of such public humiliation served as a powerful deterrent. Today's teachers may well follow a similar classroom management strategy. However, communiqués sent with wireless devices disappear once they have been sent. Thus, they change the teacher–student power dynamic because they leave the technology-disadvantaged teacher without any material evidence of wrongdoing.

Mrs. O'Neill's identity potential was clear, both to her and to us, her students. She was the detector and the squelcher of note-passing and other activities that didn't belong in the classroom. Geri Gay and Helene Hembrooke (2004), who studied "how wireless, pervasive computing affects the learning experience" of Cornell University undergrads (p. 54), explain how introducing a "wireless computer network to students to use during school hours highlight[s] the ambiguity of roles, objectives, and norms regarding tool use in the classroom" (p. 9); indeed, the Capella University ad suggests that in the presence of wireless computing, teachers cannot retain their position as arbiters of classroom interaction unless they enter the educational equivalent of AT&T's M-Life, and that to do so requires an online graduate degree in education from Capella University. Although at first glance the ad appears to support Gay and Hembrooke's contention that wireless computing changes classroom dynamics in as yet ill-understood ways, it ultimately recommends what Selber (2004) calls an "instrumental" response that supports the "status quo, relatively hierarchical student-teacher relationships" (p. 9).

I found details of Capella University's curriculum online at www.capella.edu. If the published course descriptions are to be believed, no course within the university's School of Education teaches pedagogical strategies specific to classroom integration of technology. Its benefit to teachers hoping for improved interactions with the "whole new breed of student" appears to be nothing more than the fact that it is offered online. What new responses can such a degree possibly add to a teacher's classroom management repertoire?

Figure 3.2. Unicon Ad

Two other ads and an article, all of which appeared in the array of periodicals I reviewed, may help us explore this question. The ad depicted in Figure 3.2 is sponsored by Unicon, the self-described "leading independent provider of enterprise portal technology for higher education." In the text below the image, Unicon promises to deliver "ready to use solutions for your online campus that can be easily integrated with existing services." Unicon's represented participant, "Mike," like the little girl in the Capella University ad, is a student who appears to be working alone, completely immersed in a world built of wireless technology. Mike listens to music on his headphones while "submit[ting] his homework, register[ing] for next semester, and pay[ing] his tuition through the campus portal." He works in quiet surroundings featuring a glossy table, a sofa, and a bright window. His posture is intent if ergonomically ill-advised.

Unicon delivers the following message to the interactive participants, who are, in this case, administrators rather than teachers: "Now is the time to build your online community with Academus Portal." "Now is the time" is an evocative phrase for two reasons. First, it calls up a sentence used in typing instruction, "Now is the time for all good men to come to the aid of their country," which serves to emphasize that the keyboard is the primary mode of interaction in this environment. In addition, the phrase instantiates a sense of urgency, a call for immediate action, which is reinforced by the first sentence in the smaller print: "Students expect online services that are both convenient and comprehensive."

Despite the reference to the Unicon portal as a community–building apparatus, no visible evidence of community can be found in the image. Mike is presented as a discrete entity. In this respect, he is similar to the technologies he manipulates; he and his technologies can be "seamlessly integrated" into his life and work (Selber, 2004, pp. 129-130). His head-phones shut out distractions from the outside world, while his wireless lap-top computer invisibly and unproblematically links him to all of the resources and services, both on and off campus, that he might possibly need. Mike can work anywhere. Teachers figure in the ad only obliquely; Mike is said to submit his homework through the portal, which implies that a teacher—somewhere—has assigned that homework and will eventually log in to collect it. In addition, the last paragraph of the fine print includes this recommendation to the interactive participant: "Give your students, alumni, faculty, and staff the convenience Mike enjoys—contact Unicon for a demonstration today!" Of the four stakeholder groups mentioned in the concluding statement, faculty come third, after students and alumni. Administrators' assumed priorities are clear in this ordering of stakehold-ers. Serious and seemingly typical students like Mike already engage in self-motivated learning that requires neither a teacher nor a classroom. What is the identity potential of an invisible teacher?

In Figure 3.3, a Sony Vaio ad, which appeared in *Campus Technology* in October 2004, similarly depicts students as discrete entities. Images on three Sony screens show bright, engaged students using Sony Vaios. Mike, whom we met in the Unicon ad, is here again, this time displayed on the screen of a laptop on the right side of the page.[2] A brightly smiling Asian girl clad in a yellow sweater appears on a flat-screen monitor in the center. She sits on a glossy floor typing on a small laptop computer. Like "Mike," she wears headphones. In the back-ground we can see her red bicycle, sig-nifying that her mobility is enhanced by the availability of ubiquitous, spon-taneous computing. In the third, much smaller photograph, another Sony Vaio dominates the screen of a laptop in a kind of *mise en abyme* (picture within a picture). A trendy young woman, partly hidden by the laptop's screen, works in a library amid stacks of colorful books.

Figure 3.3. Sony Vaio Ad

Thin, almost invisible angular white lines connect the three comput-er screens, thereby linking the stu-dents to one another. In front of each featured student is a drawing of a closed door, and each door faces a road, signified by a green stripe that angles through the image, drawing the viewer's eye from lower right to upper left and off into an unrealized future. Wandering along the road are shadowy cartoonish figures, who, judging from their accoutrements—bookbags, sneakers, khakis—appear to be other students. The cartoon figures' features, however, are ill-defined, especially in contrast to the crisp, colorful students depicted on the Sony monitors. Moreover, the cartoon students appear aimless, unlike the photographed students who, just beyond the closed doors, work in engaged enjoyment thanks to their Sony hardware.

The headline identifies Sony Vaio as "the technology of education," and the first sentence of the small print proclaims, "Technology has revolutionized education." In the very next sentence, the ad once again equates "technology" with "Sony": "Sony's wireless solutions extend far beyond the lecture hall, from the computer center to the administrative office" with "services optimized for education." Key words in the ad copy promise that Sony products "offer a no-compromise mobile experience," "shed new light on any subject matter," and enable "you" to "stay connected in and out of the classroom." In short, "Sony is the bright choice in education technology." In Sony's educational world, students who use wireless technologies are as bright as the Sony products they work with. They are all intent, engaged, smiling. Students who are not connected to the wireless world Sony affords are murky, aimless, and excluded by closed doors. Administrators are optimistic. Teachers are—you guessed it—absent.

The absent teacher narrative is not limited to advertising representations. Teachers are absent, as well, from the content in these publications. A case in point is the one-page article, "Look Ma, No Wires" (St. John, 2004), shown in Figure 3.4. This ad was published in the inaugural issue of *Edutopia*, a publication that bills itself (in a pop-up window on its web site) as follows: "The one magazine that will give you practical, hands-on insight into what works now, what's on the horizon, and who's shaping the changing future of education" (Edutopia). If content about wireless technology in this inaugural issue is any indication, children shape education and teachers have no recourse but to follow.

Figure 3.4. Edutopia Article

The image associated with the article depicts a little girl. Her posture and her surroundings are by now a cliché: She sits on a glossy light-colored floor in front of an overexposed window. In her red sweater, she stands out from the neutral background, and her ergonomically questionable posture, like that of "Mike" and the other students in the previous representations, conveys intense engagement. The article's title, "Look Ma, No Wires," creates an association with the classic saying, "Look Ma, no hands," which suggests mastery over an exhilarating and somewhat daring new skill, such as riding a bike "no hands." This small child is shown to be an independent learner, a bit of a risk-taker, thanks to a wireless network installed at her school. The article describes the four steps required to install a Wi-Fi network. The term "Wi-Fi" is used in the article, but the reader never learns what "Wi-Fi" stands for. The term is defined only as the "catch name for several technical standards that let you establish a high-speed connection to the Internet using a radio signal" (p. 20). The steps are as follows. (a) Have a wired connection installed. (b) Obtain Wi-Fi adapter cards for each computer to be networked. (c) Install the cards. (d) Select the channel. Sounds simple enough, but the language used in the article seems to be aimed not at the novice wireless user, but at a nonspecific mix of technicians, classroom teachers, and school media specialists. The article ends by glossing over the issue of security: "If your room is close to the outside world—say, if there's a coffee shop beyond the library wall—your neighbor may be able to pick up your signal. Odds are, though, that will be a problem only at a school in a dense urban area" (p. 20). The "Jargon Buster" sidebar includes entries for WEP and WPA, two common security standards, but these are explained in relation to their encryption levels, not in terms of school security. The final sentence reinforces the ease of the process: "[I]n virtually no time you'll have set up a simple conduit to the unlimited educational opportunities available online" (p. 20).

Wireless computing is valorized in the article, just as in the ads described earlier. Moreover, the article perpetuates the companion narrative of the absent teacher in several ways. First and most prominent of these representations is the image/headline combination, which occupies the important upper-left quadrant of the article's single page. The words, "Look Ma, No Wires," placed beside the image of the small child working alone in a seemingly empty space, argue in favor of learning without teachers. The promise of unlimited online educational opportunities with which the article concludes reinforces the implication that wireless technology is not only easy to install but can replace teachers. Finally, the article's diction and tone are exclusionary. Some technical terms are overexplained whereas others are left undefined. Procedural details are omitted, which leaves the reader with the impression that the wireless network installation process is effortless. Obviously, it is much easier to set up a network today than it was even a few short years ago, but there are still some considerations that must be addressed, including the issue of network security. In short, although the article appears in a publication that is ostensibly aimed at teachers, teachers are neither the represented participants nor the interactive participants. In fact, with image, tone, and language, the author suggests that technology—a wireless network—rather than a human—a teacher—is the bearer of educational success.

These images of the wireless world illustrate the extent to which media representations have a pedagogical function (*cf.* Jhally, 1988, p. 24; Schriver, 1997, p. 45); that is, they propose to readers an identity potential—a repertoire of behaviors, attitudes, and values available within readers' disciplinary and personal domains (Kimme Hea, 2002; Kitalong, 2000, 2004; Selfe, 1999). The absence of teachers as represented participants in media representations of academic wireless technology use affords teachers an impoverished identity potential with respect to technology. The invisible teacher narrative is just one element in a prominent and comprehensive suite of narratives that pervade today's educational discourse. These persistent "companion narratives" (Kimme Hea, 2002; Selfe, 1999) together suggest that teachers are self-evidently and emphatically much less technologically competent than their students and insinuate that because teachers can never catch up with their students' technological abilities, they should not even bother to try. In short, although the ads and articles are seductive in the simplicity of their instrumental solutions, we must be realistic: No teacher can fully understand the complex wireless world by enrolling in Capella University or any other expensive private online curriculum, especially one that apparently includes no direct technology instruction of any kind. And no school can ensure the building of community and student engagement simply by instituting technological solutions such as portals, wireless networks, or Sony monitors.

The invisibility of teachers in ads for wireless technology should not be surprising, when we consider the multiple, intertwined ways in which teachers are always already invisible in the academy. Every teacher has encountered students who purport not to remember our names, even at midsemester. Moreover, many teachers labor as contingent workers, lacking basic amenities such as offices, parking spaces, health insurance, union membership, or even a living wage. Teaching on line, although attractive to some and becoming increasingly common throughout the academy, exacerbates the invisibility problem by erasing teachers' and students' faces along with the timbre of their voices, their mannerisms, and in some cases, even their personalities.

It is my contention, however, that teachers need not accept their invisibility. Narratives like the ones analyzed in this chapter are open to reinterpretation and rearticulation. In the remaining pages of this chapter, I explore how a three-part taxonomy of computer literacies described by Stuart Selber (2004) can be mobilized to help teachers and students analyze, understand, and ultimately shape what Selber calls the "conceptual landscape" of computer literacies (2004, p. 24), which, of course, includes the wireless world.

In *Multiliteracies for a Digital Age,* Selber outlines three categories of technological literacy—functional, critical, and rhetorical (Selber, 2004, p. 24). Paying attention to all three categories[3] at once allows students—and for the purposes of this chapter, teachers—to "participate fully and meaningfully in technological activities" (Selber, 2004, p. 25). I use Selber's framework to suggest strategies teachers can use to rearticulate their own identity potential and thereby counter how they are portrayed in media representations of wireless technology. Selber's framework is useful because each category of literacy in his three-part taxonomy is associated with a distinct subject position. Armed with the broadened identity potential made possible by a more robust understanding of the technological landscape, technology users, both teachers and students, can better identify and ultimately "exploit the different subjectivities that have become associated with computer technologies" (Selber, 2004, p. 25).

Selber's taxonomy does not dismiss "mere" functional literacy; on the contrary, each category in his taxonomy is viewed as important as the others. The three categories reinforce and augment one another based on the assumption that no single category of technological literacy is deemed sufficient for full participation in the technological landscape. Students in Selber's argument, and teachers in mine, must be "adequately exposed" to the entire conceptual landscape, not limited to one or two of the three possible and interrelated ways of understanding technological literacy (Selber, 2004, pp. 24-25).

Selber associates *functional literacy* with the use of computers as tools. The subject position he associates with functional literacy is the technology user, and the ultimate objective of functional literacy is the ability to parlay technology skill into effective employment (Selber, 2004, p. 25). By itself, functional literacy offers an impoverished identity potential for teachers and students, because it constructs technology use as decontextualized. This means that, for example, every teacher's classroom technology challenges would be addressed by having her learn additional software or hardware tools.

A functional literacy approach underlies the Capella University ad's message: It construes social PDA use as a classroom problem and posits a master's degree in education as the solution. The ad taps into teachers' insecurities that technologically adept students will circumvent the teachers' rightful authority. Selber argues that functional literacy is necessary but not sufficient. He strongly advocates a solid technological functional literacy, but believes that it is inadvisable for technological literacy to serve as the end point.

Selber's second category, critical literacy, moves beyond functional literacy to conceptualize technologies as cultural artifacts and their users as questioners of technology, in particular of the "politics of technology" (Selber, 2004, pp. 25, 75). The ultimate goal of critical technological literacy is an informed critique of technology, during which the critic interrogates the *status quo* of technology development and use through questions such as the following (Selber, 2004, p. 81):

- What is lost? What is gained?
- Who profits?
- Who is left behind and why?
- What is privileged in terms of literacy, learning, and cultural capital?
- What political and cultural values are embedded within the hardware or software?

Using questions such as these to critique the messages embedded within the Capella University ad, teachers might ask, for example, which of their students own and use PDAs, cell phones, and laptop computers, and which students do not. Together with their students, they could speculate about what is lost and gained with ubiquitous wireless technology: How do wireless technologies change social interactions both inside and outside the classroom?

Teachers who ask critical questions, and who engage students in exploring such questions, expand their own identity potential in that they are led to consider a range of possible responses to wireless technologies. Teachers who adopt a critical literacy emphasis would be more likely to explore whether wireless technologies can function as something other than classroom intrusions; in other words, whether they might offer any literacy and learning benefits.

A key aspect of the critical literacy approach to analyzing wireless technology is that the artifact under study must be contextualized within the larger cultural landscape. It is easy to assume, operating from a functional literacy perspective, that technologies and technology users can be "understood independently of larger social structures and forces," that they are discrete entities (Selber, 2004, p. 129), that a wireless apparatus used in an elementary school classroom has no connection to the wireless world at large. The invisible teacher in the Capella University ad can develop a critical understanding of the PDA-using note-passer as enmeshed within an expanding network of wireless technologies. Similarly, adopting a critical literacy perspective toward the Sony ad opens up questions about the desired relationship between the clear, bright students equipped with Sony products and their shadowy cartoon counterparts wandering aimlessly along a road to some unspecified future. Does equipment alone ensure that students will fulfill their intellectual potential? Of course not. But without appropriate technology coupled with the wherewithal to use it productively, the pathway is unclear, the outcomes uncertain.

Enter Selber's final category, rhetorical literacy. In his vision, functional and critical technological literacy are necessary aspects of technological literacy, but true technological literacy occurs only when students are equipped to engage in social action—that is, when they adopt a subject position that allows them to become the architects of their own online environments or interfaces. By facilitating their own and their students' rhetorical literacy, teachers can begin to "recognize the persuasive dimensions of human-computer interfaces and the deliberative and reflective aspects of interface design," which make it possible to effect "social action" (Selber, 2004, p. 140).

The computer as it is associated with rhetorical literacy is conceptualized as a hypertextual medium, in that it is made up of "(nonlinear) text, (modular) nodes, and (associative) links" (Selber, 2004, p. 168). Because of the interactivity hypertext entails, even simply reading an online document can be viewed as a form of production. When users click on links and choose paths through an online document, they "write their own versions of [the] text . . ." (Bolter, qtd. in Selber, 2004, p. 169), thereby constructing a seemingly unique experience upon each reading. To achieve rhetorical literacy, however, is not as straightforward as constructing hypertextual readings of online texts. Indeed, such construction could be, at base, an utterly passive experience akin to channel surfing. Rhetorical literacy entails, in addition, the development of a critical awareness of one's own and others' productive power through a process of reflective praxis, the "thoughtful integration of functional and critical abilities in the design and evaluation of computer interfaces" (Selber, 2004, p. 145). In Selber's framework (2004, p. 147), someone who exhibits a rhetorical approach to technological literacy:

- Understands that technology development and use are persuasive practices that are contextualized within larger contexts and forces;
- Understands that problems associated with the design, development, and uses of technology are often ill defined and therefore require sustained deliberation and experimentation;
- Can consciously articulate what he or she knows about technology development and use;
- Critically assesses and reflects upon this articulated knowledge;
- Sees technology development and use not as technical action but as a form of social action.

The ads presented in this chapter do nothing to foster critical or rhetorical literacy. In fact, one might argue that ads and other media representations actively discourage readers' critical or rhetorical awareness by offering simplistic technical solutions to complex social issues, as this chapter's discussion illustrates. How, then, can teachers move from a functional to a critical response or rhetorical response when faced with promotions such as the Capella University ad? How can students and teachers be compelled to rhetorical literacy? Although the Capella University ad depicts an elementary school child using a cultural artifact to produce and transmit written communication, it does not serve as encouragement to rhetorical literacy. In fact, it dissuades ideological questioning and actively discourages the productive redesign of the classroom technological environment.

Capella University sends a definitive functional literacy message: Teachers should enroll in this particular graduate program to master technology. In so doing, they will become more effective teachers and more indispensable employees, a functional literacy goal. The teacher's employment effectiveness will apparently be judged by her ability to detect and squelch students' technological productivity. This recommended response solidifies the classroom as the primary site of learning, the teacher as maintainer of classroom decorum, and the student as someone whose rhetorical literacy—the remediation of an existing communication practice—must be squelched.

When we interpret this ad from the vantage point of rhetorical literacy, however, we are faced with the prospect of translating what we know about technology into opportunities to design and implement new, more robust technological landscapes. "In order to function most effectively as agents of change," Selber writes, we "must also become reflective producers of technology, a role that involves a combination of functional and critical abilities" (2004, p. 182). Acknowledging that "interface design is a brave new world for many humanists," he suggests constructing it as a "rhetorical activity . . . that includes persuasion, deliberation, reflection, social action, and an ability to analyze metaphors" (2004, p. 182). For teachers, this may mean producing a classroom interface that expands to embrace the use of wireless technologies for research, communication, and other extended learning activities.

If we believe, as I have argued here, that the pedagogical function of media representations is a persuasive cultural force that significantly shapes teachers' identity potential, then it is incumbent upon teachers to challenge such media representations both within and outside of their classrooms. "[T]hose who are rhetorically literate," Selber reminds us, "understand that persuasion always involves larger structures and forces" (2004, p. 150).

I have tried to illustrate in this chapter the technological, institutional, social, economic, and political structures and forces that conspire to render teachers absent to students. And teachers may feel powerless to change those insistent ideological forces. But Selber suggests a remedy: If teachers become rhetorically literate, they place themselves "in a unique position to design agreeable and worthwhile interfaces" (2004, p. 150). These interfaces, can, of course, be of the screen-based kind—hypertextual representations of knowledge, emotion, and engagement that encourage thoughtful reflection and interactivity. But interfaces can also be more three-dimensional, live, face-to-face. Through rhetorical literacy, teachers can, in short, create themselves as interfaces, as owners of a fully functional, robust identity potential.

The presence of wireless technology changes existing cultural practices. Wireless technology necessarily permeates the classroom, where it challenges power structures, generates new practices, and begets hybrid learning contexts. Teachers, like students, can revert to the tried-and-true functional responses to such changes—the same responses that rendered teachers invisible in the first place. Or they can choose to produce, by practicing rhetorical literacy, a robust, questioning, and productive classroom interface that includes a vigorous and multifaceted identity potential for teachers. Although the identity potential enabled by the advertising image of a little girl writing a technology-mediated love note has seemingly been "fixed" (Burgin, 1999, p. 49) by textual associations with online education and the companion narratives of teacher invisibility and student expertise, it need not be "contained" (Burgin, 1999, p. 49) by these textual associations. In classrooms around the country, then, teachers and students need to develop functional literacy by familiarizing themselves with everyday wireless technologies, develop critical literacy by questioning the assumptions that accompany media representations of wireless technologies, and develop rhetorical literacy by incorporating such technologies as tools for extending the classroom in meaningful ways. In so doing, teachers may begin the task of severing image/text bonds that prematurely exhaust their identity potential.

ENDNOTES

1. The term "identity potential" is adapted from Kress and Van Leeuwen's (1996) notion of meaning potential, which they define as "what you can mean and how you can 'say' it" (p. 8).
2. The fact that Sony and Unicon used the exact same image mystified me until I began to research stock photography databases for another project and learned the extent to which advertising agencies, news outlets, and content providers acquire generic images from such repositories. (See Frosh, 2003; Kitalong, 2005.)
3. Some might prefer the term levels instead of categories, but in fact, simultaneous attention to all three literacies is crucial. They cannot, in my estimation, be attended to sequentially or arranged on a continuum or timeline. I believe that Selber would agree.

REFERENCES

Banner ads tenth birthday. (2004, October 27). *Adland.* Retrieved November 5, 2005, from http://ad-rag.com/114815.php

Burgin, Victor. (1999). Art, common sense, and photography. In Jessica Evans & Stuart Hall (Eds.), *Visual culture: The reader* (pp. 41-50). London: Sage.

Edutopia. (2005). Retrieved November 5, 2005, from http://www.edutopia.org/index.php

Friedman, Thomas L. (2005). *The world is flat: A brief history of the twenty-first century.* New York: Farrar, Strauss, Giroux.

Frosh, Paul. (2003). *The image factory: Consumer culture, photography and the visual content industry.* Oxford, England: Berg Publishers.

Gay, Geri, & Hembrooke, Helene. (2004). *Activity-centered design: An ecological approach to designing smart tools and usable systems.* Cambridge, MA: MIT Press.

Jhally, Sut. (1988). *The codes of advertising: Fetishism and the political economy of meaning in the consumer society.* New York: Routledge.

Kimme Hea, Amy C. (2002). Rearticulating e-dentities in the web-based classroom: One technoresearcher's exploration of power & the WWW. *Computers and Composition, 19,* 331-346.

Kitalong, Karla Saari. (2000, July). 'You will': Technology, magic, and the cultural contexts of technical communication. *Journal of Business and Technical Communication, 14,* 289-314.

Kitalong, Karla Saari. (2004). Who are the users? Media representations as audience-analysis teaching tools. In Tracy Bridgeford, Karla Saari Kitalong, & Dickie Selfe (Eds.), *Innovative approaches to teaching technical communication* (pp.168-182). Logan: Utah State University Press.

Kitalong, Karla Saari. (2005, October). An anti-feminist rhetoric of stock photography? Digital asset management, metatags, and the consequences of classification. Paper presented at Feminisms and Rhetorics Conference, Houghton, MI.

Kress, Gunther, & van Leeuwen, Theo. (1996). *Reading images: The grammar of visual design.* London: Routledge.

Schriver, Karen. (1997). *Document design: Creating texts for readers.* New York: Wiley.

Selber, Stuart A. (2004). *Multiliteracies for a digital age.* Carbondale: Southern Illinois University Press.

Selfe, Cynthia L. (1999). Lest we think the revolution is a revolution: Images of technology and the nature of change. In Gail E. Hawisher & Cynthia L. Selfe (Eds.), *Passions, pedagogies, and 21st century technologies* (pp. 292-322). Logan: Utah State University Press.

St. John, Donald. (2004, September/October). Look ma, no wires. *Edutopia,* Issue 1, p. 20. Retrieved August 17, 2005, from http://www.edutopia.org/magazine/ed1article.php?id = art_1154&issue = sept_04#

4

REWRITING WI-FI

THE SURVEILLANCE OF MOBILITY
AND STUDENT AGENCY[1]

Ryan M. Moeller

Although it is relatively new to university campuses, wi-fi technology[2] is already claiming new paradigms of the mobile worker, the mobile workplace, and even the mobile shopper. In industry terms, mobility typically means freeing such technologies as computers, personal digital assistants (PDAs), entertainment hardware, and telephones from wires, from land lines, from government licensing, and from telecommunications giants. The promises are great, but so are the caveats: The paradigms of mobility are conceived and articulated under metaphors of increased access and efficiency and decreased downtime but often result in increased surveillance and accountability (Brans, 2003; Easton, 2002; Hogan, 2001). I do not believe it's a stretch to assume that these same paradigms will affect students who are subjected to (and who subject themselves to) wireless technologies in the writing classroom. Bryan Alexander (2004) details some ways in which mobile technologies are influencing teacher and student behaviors with regard to the classroom space (p. 29). For Alexander, the difference between "wired" and "mobile" technologies is that the latter are more personal, more intimate technologies that are "held close to the body—in a purse, on the lap, in a pocket, on the floor next to the user. Their screens are easily hidden from prying eyes. Emotional investments increase, even with shared devices" (p. 30). In light of this difference, Alexander advocates pedagogies that follow a new, emotionally invested, "nomadic" learning trend (p. 34).

Like the paradigms of mobility, there are caveats to nomadic pedagogies. Mahesh Bhave (2002) argues that the difficulty with wireless technologies is that they are disruptive as well as learning-enhancing: "Phones, smart phones and connected PDAs have already rendered the classroom permeable to outside interruptions. . . . Simpler protocols, such as requiring students to turn off their phones or set them to vibrate mode, are no longer sufficient" (Conclusions section, last node, ¶ 2). What this means for composition studies is that wi-fi technologies come with obvious beneficial capabilities: The extension of the classroom and research site to anywhere on campus at any time, the ability for several computers (and thus several students) to share a single internet connection, and increased collaboration among students using different technologies. But like the mobility offered to workers and consumers, the realities of the mobility offered to students come at a price. I find it likely that the incorporation of mobile technologies into the writing classroom will parallel their place in the business world: Increased mobility will include greater surveillance of students, their work, and their play. Due to network security concerns, administrators will require students to be more accountable for *working* more and *playing* less in the classroom. This makes sense, because the demand for teachers to prepare students to be a part of a new, mobile workforce will likely increase.

As Christine Haas and Christine Neuwirth (1994) point out, technologies almost always have an effect on writing practices in the classroom (p. 320). Computer technologies often reduce planning and drafting, for example (p. 322). And, as Gail E. Hawisher and Cynthia L. Selfe (1991) show in their landmark article, these effects are not always positive. In fact, Hawisher and Selfe argue that networked computer access to student writing may have a panoptic, disciplinary effect on students' writing practices (pp. 56, 62-63). Wireless access could certainly heighten this effect, and it already has in many respects. Joseph Janangelo (1991) advocates for an "awareness of the computer's capacity to increase human exploitation, and a knowledge of how it can constrain, complicate, and compromise our ability to teach effectively" (Conclusion section, ¶ 1), especially when computer technologies allow teachers to "monitor their students' lab attendance, lab hours, and composing time" (¶ 3). One of Janangelo's critiques is that teachers will often track student logins as a disciplinary measure when work is required outside of class and facilitated through networked technologies (¶ 3-7). Wireless technologies do not alleviate such options as spying on students' writing activities. In fact, wireless technologies may make such practices easier. Recently, MIT deployed a wireless tracking network that provides exact numbers of wireless users at any given location, and even personal information if users have opted to make it available.

The promises of wi-fi are detailed in product packaging and advertisements, and it is these promises that motivate teachers, students, employers, and workers to adopt mobile technologies into their writing and working practices. Karla Saari Kitalong (2000) identifies three reasons why technical communicators, at least, should pay attention to technology advertisements: Advertisements often portray technologies and technical processes as being "magical" or as possessing transformative powers that go unquestioned through the discourse of advertising; advertisements represent the most "widespread source of technological information" that often "create and sustain consumers' optimistic expectations" for new technologies; and technical communicators can often unwittingly support "magical" discourses of technologies through their own practices (pp. 289-291). Drawing on advertisements for wireless and mobile devices and services, I argue that writing teachers should work to make students aware of the ways in which mobility and mobile technologies are affecting their writing practices. Sustained research *with* students into the ways that mobile technologies affect writing practices will demonstrate that the powerful message of advertising is not "decidedly peripheral to the real work of society" (Kitalong, p. 295) and expose the sociopolitical implications of technological products used in the classroom (Haas & Neuwirth, 1994; Hawisher & Selfe, 1991; Janangelo, 1991).

To demonstrate the ways that advertising attempts to influence potential wi-fi users, I investigate the dominant rhetorical appeals of wi-fi products that are playing a crucial role in the deployment of wireless networking technology. First, by appealing to employers and employees separately, the wi-fi industry offers solutions to *work* that promises to make it less like work while promising greater productivity and surveillance of employee work practices. Second, by appealing to home entertainment, the wi-fi industry further conflates the difference between work and play while tracking consumer practices. Third, in their appeals to security, the wi-fi industry scares consumers into protecting their data and their broadband connections, using that fear to sell more products. Finally, by appealing to the unlicensed frequency on which wi-fi is broadcast, the wi-fi industry promises greater consumer choice and freedom from government regulation while subjecting consumers to industry-level regulatory practices. I argue that students and instructors in the writing classroom can critically understand these appeals by engaging them through rhetorical analysis as well as qualitative and sociological research into wi-fi deployment. These strategies fall under the scope of social-epistemic rhetoric, or the active, dialectical inquiry into the political economies of wireless technologies within the space of the writing classroom.

RHETORICAL APPEAL: REFIGURING WORK

Mobility advertisements depict workers being productive outside the dreary cubicle setting, people maintaining communications with one another in contexts where such communication is not typically possible (airports, parks, etc.), and friends using the Internet from comfortable coffee-house couches. The primary appeal of mobility, however, is to employers who seek greater productivity from workers. Marketers promise employers increased work time from their employees—especially those who travel—and greater compatibility between current communications infrastructures and wi-fi networks. But the recent surge in wi-fi technologies is about more than just new networks. Patrick Hogan (2001) reports, "Wireless networks have been around for years but have never really caught on. Part of the problem was their sub-2Mbps transfer speeds. More than that . . . there has never been the kind of marketing push and distribution infrastructures—not to mention technical support—that are behind wi-fi today" (Wired Vs. Wireless section, node 3, ¶ 5). Essentially, marketing and distribution have made sure that many *employers* are aware of the promising potential of wi-fi as a part of their business and network infrastructures.

To this effect, Intel Corporation cites internal studies that prove wireless technologies aid employee productivity:

> With Intel Centrino mobile technology, you can give employees the freedom to work wherever it's convenient, increasing collaboration and job satisfaction. . . . Coupled with a wireless LAN solution, your company reaps the benefits of reduced infrastructure costs, along with greater adaptability and expandability as business needs change over time.[. . .] In a recent study, Intel IT found that workers who used wireless notebook PCs with Intel Centrino mobile technology gained more than two hours per week of extra productivity. ("Increase," n.d.)

This marketing technique—aimed at employers through the careful use of the second person—exploits an employer's familiarity with cost-benefit analyses to warrant the initial expense of purchasing wi-fi technology. Increased productivity is bolstered through convenience, collaboration, and satisfaction.

Another type of appeal, aimed at employees rather than employers, attempts to garner support for wi-fi consumption from within businesses. One of the key promises behind this appeal is the illusion that work is not work, especially if it does not take place in traditional *workplace* settings. Kitalong (2000) identifies this melding of work and leisure as a part of the "magic" embodied within technology advertising (p. 295). A prominent IBM Thinkpad advertisement asserts: "You're not chained to your desk. You're wired to it." In this case, technologies—not work or employers—constrain workers and limit productivity. Greater productivity and success can be achieved easily by simply being "unwired" from the workplace. These appeals to mobility in the workplace promote a wide-scale adoption of wi-fi technologies.

Outside of workplace constructions of mobility, the wireless industry presents a limited view of mobility. Nokia, a leader in wireless technologies particularly noteworthy for its distributed models of research and infrastructure,[3] has an extremely restricted view of its users. Mats Lindgren, Jörgen Jedbratt, and Erika Svensson (2001) present Nokia's conceptualizations of three consumer groups: "*teens* (up to 18), *students* (19-25), [and] *young professionals*" who demonstrate only "slightly different needs" (p. 118). Such market characterizations, coupled with mass marketing strategies, lead Lindgren, Jedbratt, and Svensson to predict "micro-geographical marketing" or mobile advertising will be broadcast to the wi-fi user based upon location rather than specified, individualized needs (pp. 125-127). These advertisements do not consider activities outside of consumerism; rather, they reposition all other activities as a part of consumerism. The result, then, of the limitations that the wireless industry places on wi-fi—whether they reconceptualize the workplace as being *everywhere* or consumerism as a primary human activity—is that they severely limit users' agency.

When targeting university administrators, advertisers emphasize increased collaboration among faculty and students, reduced infrastructure and technical costs, and increased productivity. For example, Tangent is a company that specializes in selling wireless laptop carts. Their advertising brochure printed in 2002 makes the following promises:

> One wireless lab can become ten or more classrooms, for as many teachers. Budgeted funds and grants go farther, enabling administrators to bring the benefits of technology to more students. . . . Wireless classrooms can also reduce infrastructure and maintenance costs associated with hard-wiring a school network for communications and the Internet, in some cases virtually eliminating class to class cabling. (p. 1)

These promises claim that wi-fi allows technology to extend further into the learning process and students' lives with fewer costs and technical hurdles—cabling, budgeting, and even the digital divide—by enabling "more" students to have access to technology. Intel's own advertising to university students parallels this promise of access and collaboration: "[C]olleges are setting up wireless LANs and hotspots that give students and faculty with mobile notebooks immediate access to people and data from anywhere on campus . . . greatly expanding the ability of the entire community to collate, analyze, and synthesize information and transform it into knowledge" ("Living Wireless," n.d.). Despite the sense of urgency created by the advertisements, there are many concerns, voiced by Alexander (2004), Bhave (2002), and others, that this new mobility might prove more disruptive than productive. In the classroom, increased surveillance and teacher control will likely evolve not from the mobility of the technology but from the many other functions that serve as distractions in a classroom environment. For the purposes of this argument and because the industry already does so, I classify these "distractions" under the broader category, *entertainment*, and they lead to increased security and policing of wireless technologies in classrooms and on university campuses.

RHETORICAL APPEAL: MAKING ENTERTAINMENT PRODUCTIVE

Most critiques of wi-fi in the classroom focus on the disruptive nature of the technology, the distractions that hand-held and notebook computers provide to users. Certainly, there have always been opportunities for students who are easily distracted: passing notes in class, doodling, and day-dreaming, among them. The difference with wi-fi technologies is in their capacity to merge activities which have been historically separated in time and space: entertainment and work. Wi-fi television, control of home entertainment systems, broadcasting of music files and streaming videos from computers, and home automation are among the entertainment options now available to consumers. These options promise a convergence of entertainment technologies under a single technology such as a hand-held or a notebook computer. With just a few add-on technologies and a computer, a user may listen to music or play a movie downloaded from the Internet on her television in any room of the house or on the classroom projector. She could watch a cable or digital television program anywhere without the need for a hardwire connection. Already, many wi-fi enabled devices such as computers or PDAs serve multiple functions as calendar schedulers, contact list managers, MP3 players, and even remote controls. For example, many HEWLETT PACKARD computers can now be controlled by remote control, demonstrating a convergence between entertainment and productivity. Intel is working on an interface that treats the computer primarily as a wireless access point for controlling television and music programming (Gohring, 2004). Pocket PCs have optional software allowing them act as a universal remote control for wi-fi or Bluetooth enabled products such as CD players, televisions, or DVD players. Bluetooth–enabled MP3 music players can download music files queued from a computer database.

Several innovations in wi-fi products have been developed primarily for entertainment value—that is, they serve interests that are not typically associated with work. Joseph Palenchar (2001) hopes that such developments will spur additional growth in wi-fi sales. In "Enhancements to Aid CE Sales," he writes that an optional 802.11e standard "will support streaming audio and video" and will "lay the groundwork for consumer electronics companies to offer personal video recorders, residential gateways, mobile web pads, and digital audio and video jukeboxes" (p. 24). The result of all of these innovations, argues market research analyst and consultant Elizabeth Parks (2004), is the development of the computer as the central multimedia platform (¶ 4). She cites as evidence the fact that 40% of computer sales in the last five years have been "media PCs," or computers designed for handling multimedia content (Parks, 2004, ¶ 1).

As Alex Lightman and William Rojas (2002) have argued, such developments would render the interoperability of the required networks and subsequently the Internet ubiquitous, making tracking consumer preferences and behaviors extremely easy and equally ubiquitous (p. 75). Such ubiquity would aid marketers in choicing[4] new products and services through consumer behavior tracking. Advertised as a value-added feature to the consumer, as in the case of saving "favorite" settings, such market research is nearly invisible and totally efficient. Data could be transferred easily from a consumer's network to the product marketers on a regular basis. This strategy is already employed in many computers that will automatically generate and send reports to the hardware or software manufacturer when errors occur. As Robin Mansell and W. Edward Steinmueller (2000) warn about in terms of "broadcast models" of distribution, such simple choices as whether or not to send a report or whether or not to save one's preferences represent a wide body of negotiation between regulators, product developers, and marketers that are obscured at the level of consumption (p. 66). As the consumer electronics products that will organize and distribute our *entertainment* and *leisure* through wi-fi are developed, consumers will influence the industry passively by letting it track their individual use.

The most problematic fact about this convergence of entertainment technologies under the wi-fi protocol—at least as far as its impact in the writing classroom—is that it is done under the name of entertainment. Neil Postman (1985) argues that entertainment, and primarily television entertainment, is the vehicle through which U.S. culture identifies itself (pp. 78-80). Because of the pervasiveness of this medium, U.S. citizens see most cultural events through the genre of entertainment, even when these events are serious and their implications may be disastrous. When consumer electronics marketers appeal to such blurred concepts as *serious play* or urge consumers to *play hard* and *unwire your life*, they are drawing upon deeply embedded cultural assumptions. These assumptions place entertainment in binary opposition to work and suggest that through consumer electronics products, work can be entertaining. The problem for many university instructors is that the work-play contradiction calls into question the value of the class time. Ian Ayres states the problem: "When you see 25 percent of the screens playing solitaire, besides its being distracting, you feel like a sucker for paying attention" (qtd. in Heller, 2003, ¶ 1). At the risk of coming across as a critic of play as a productive learning tool, I want to emphasize that wi-fi products are designed to make work seem more like entertainment and to make entertainment activities more productive by tracking consumer behaviors. This places unnecessary pressure on instructors to do the same—to make class activities more fun *and* more productive at the same time—by virtue of competing for students' attention. Additionally, if assessment can be done on consumers without their knowledge, what implications are there for instructors to do the same? Although this development supports some instructors' pedagogical goals, it places unilateral pressure on all instructors to *entertain* as well as to teach and assess student learning. This would require a complete paradigm shift in current educational practice that seems costly at best.

RHETORICAL APPEAL: SELLING SECURITY

Although many instructors are working to change their course delivery methods to accommodate connectivity and mobility and to create new ways to keep their students' attention (*cf.* Alexander [2004] and Prensky [2005]), other factors make wi-fi accessibility in classrooms problematic. For technical support personnel on university campuses across the country, security and network integrity is foremost on their minds. When I first asked about wireless connectivity at my institution, I was encouraged to wait. Of primary concern, I discovered, was network security and maintaining control over everyone connected to the network. I was told about hackers—one infamous French hacker in particular—who were looking for open access points left by faculty or staff who compromised the network by installing their own, renegade wireless routers. This concern is validated in the product packaging of wi-fi routers that devotes much valuable packaging space to discussions of security and privacy.

The 2002 packaging of D-Link's DI-614 + wi-fi router claims that the primary purpose of such a technology is to "Protect and Share" a connection to the Internet. In fact, on the front of the box alone, there are four separate appeals to security: *protection, firewall, parental control,* and *data encryption.* One of the many messages implicit in this repetition of security themes is that there must be something to fear from broadcasting or sharing an Internet connection. In *The Culture of Fear,* Barry Glassner (1999) points out that the Internet is often touted by the mainstream press as a "city without cops," or a safe haven for child pornographers and pedophiles (pp. 34-35). Glassner's book refers to what Catherine Chaput (2004) has called a "political economy of fear" (n.p.). Manifested in reality television shows, disproportionate media coverage of violence and criminal behavior, planned and guard-gated communities, and news media coverage of fear-induced consumerism (scenes at shopping centers hours before a hurricane, for example), fear produces tangible results for marketers and television networks. Communications scholar John Fiske (1996) argues that a constant barrage of fear-inducing images has produced an "enclave society" in which people sequester themselves in their homes and communities (p. 242).

Chaput (2004) and Fiske (1996) both point to increased surveillance techniques as both source and product of the political economy of fear. Fear almost always justifies the use of more and greater surveillance techniques. In *Weapons of Mass Deception,* Sheldon Rampton and John Stauber (2003) link the perpetuation of fear in mainstream U.S. culture to increased consumerism. They point to the commercialization of military hardware such as a Humvee army vehicle into a high-end, all-terrain vehicle or the marketing of SUVs as "urban assault luxury vehicles" that protect consumers from the dangers inherent to driving as indicative of marketers' predilection to take advantage of consumers' fears (p. 135). In the days following the 9/11 attacks on the World Trade Center in New York, when the entire country seemed gripped with fear and grief, President Bush urged citizens to do their patriotic duty and shop. Such a suggestion provides a concrete link between fear that is incited by mass media coverage of world events and consumerism.

Similarly, wi-fi products are marketed through metaphors of security, protection, and seclusion. Well before the 2001 terrorist attacks, Charles Mason (1998) argued that "As the cellular industry has found, many customers have been attracted to wireless voice services due to security concerns" (p. 72). Mason posits that customers want wireless technologies to "automatically notify authorities of an accident, and guide them to the car; track stolen vehicles; provide navigation assistance to lost drivers; call emergency roadside assistance; and perform remote diagnostics of engine functions" (p. 72). Currently, OnStar provides such features. Sold as a subscription service in many automobiles manufactured in the U.S., OnStar operators are available via hands-free wireless connections. These operators can perform services as mundane as announcing movie times and as important as directing emergency services to the scene of an accident. It is the *important* tasks that OnStar uses in its advertising campaigns, demonstrating the value of its services to a culture in fear of the dangers of driving.

Fear of identity theft and computer hacking currently drive the wi-fi industry's product security features and much of universities' uses of wi-fi— at least at my institution. Corporate and institutional concerns over data and transaction security have led to the development of Virtual Private Networks (VPNs) and wi-fi encryption technologies that block unwanted users from accessing these networks. Fears prompted by computer users with "hotspot locators" or wi-fi sniffing devices have prompted the wi-fi industry to offer additional security measures to keep these would-be predators off of any bandwidth that might extend beyond the walls of company headquarters or outside the campus. Reports of identity theft and credit card warnings to protect one's electronic transactions help persuade consumers that they should protect their identities and their information from wi-fi predators. As seen in the appeals to security on the D-Link product packaging, marketers build these fears and then effectively offer to assuage them.

RHETORICAL APPEAL: DEREGULATION

One of the ways in which the U.S. legislature has attempted to counter public fear of wi-fi is through the appearance of deregulating the industry in order to give individuals a greater range of choice. However, the legislation leading to deregulation will not solve the problems wi-fi raises, because deregulation takes many choices away from the consumer. Chris Fisher (2002) summarizes the industry's position on government regulation:

[A] key event took place in February when the Federal Communications Commission (FCC) approved unlicensed spectrum for commercial ultra-wideband use. . . . Opening up this free spectrum gives technology companies many more options in developing wireless connectivity solutions and, in turn, gives consumer electronics companies more options to create differentiating features for their products. (p. 16)

This is the "trickle-down" theory of technology development. The logic is that greater options for consumer electronics companies will mean greater choice for consumers. Consumer choice will—in turn—drive the industry. It is important to note, however, that the consumer is not mentioned in Fisher's view of a deregulated ultra-wideband spectrum. Such a claim that ignores the consumer, the supposed instigator and regulator of the industry, is indicative of legislative discussions of products, product offerings, and regulation. In these discussions, the industry drives itself, and consumers accept and readily buy technologies that are offered to them.

In the policy hearings on broadband mergers and high-speed Internet access, consumer choice seems to be confused with *competition*. In the 2001 U.S. Senate Judiciary hearings on the America On-Line (AOL)/Time Warner merger, Senator Orrin Hatch argued that "Competition is essential to both the future of the Internet and continued innovation in the high-technology world" (p. 1). At one point in the hearing, though, representatives from AOL were questioned about consumer complaints that installing AOL's operating software erases any other ISP's software previously installed on their computers. This erasure, complainants charged, is not explained in AOL's installation program. The AOL representatives responded that, "we do tell people what we are doing. We are [erasing records of other service providers] because it is better for consumers, and indeed for 95 percent of our members they only use AOL" (pp. 49-50). Here, consumer choice appears to be a highly manipulated concept. AOL obscures their customers' possible choices of other ISPs through technical language (for example, "default ISP") in the installation process. Because a majority of users may not know what an ISP is in the first place, their choice is obscured. The spirit of competition is moot through the "default settings" of the installation software. Until recently, AOL maintained its own interface and information retrieval system that left many users thinking that AOL owned much of the content it filtered directly from the Internet. Microsoft Network (MSN) has a similar browser that sets up its own Internet web site, search engine, and news service as a default. Of course, advanced users can change default settings, but these changes are hidden behind default options set up when first installing the software. The result is that subscribers will likely use only the services selected for them rather than competing services. Users of MSN, for example, will tend to engage news, advertisements, and offers made by Microsoft more often than those from a different ISP.

Rather than policing through content control, universities such as mine police wireless usage through much more obvious measures. At Utah State University (USU), every internet connection is monitored in real time and displayed on a large network map. Because much of the network distribution is decentralized and controlled by the various colleges and departments around campus, network administrators monitor all activity and kill any port with suspicious behavior. This activity is posted on a public site. The effect of this constant monitoring and public announcement of *killed* ports enacts the disciplinary function of the institution (Hawisher & Selfe, 1991; Janangelo, 1991). Like the legislative discussions on regulating internet broadband and wireless access, the USU's own network policing demonstrates little concern for the actions of individual users except to the extent to which they threaten the functionality of the larger system. This lack of concern reaches the students through various contact points in the classroom—managing access to the network, blocking certain features, increasing workloads from having to login to class discussions outside of class time—placing more responsibility on the teacher to curb the negative behaviors of students in the class.

SOCIAL EPISTEMIC RHETORIC AND REWRITING WI-FI PROLIFERATION

These appeals to mobility, entertainment, and security instruct students to assume that wi-fi products will

- liberate them from bondage in the classroom while subjecting them to increased surveillance and accountability during leisure hours;
- protect them from fears of identity theft or susceptibility to hackers by getting them to secure their own internet connections or broadband access or by giving network administrators access to private information; and
- entertain them and assist them in negotiating their leisure time, further conflating the distinction between work and leisure within the classroom space.

As we composition instructors and administrators begin to incorporate wi-fi technologies into our writing classrooms, we must struggle to question marketing constructions of wi-fi and to maximize the liberatory potentials of these technologies: open spectrum frequencies, decentralized and user-based networks, and community access. James Berlin's (1996) social-epistemic rhetoric offers potential for integrating wi-fi technology in the writing classroom by making the contradictions of wi-fi technology outlined above—mobility/surveillance, entertainment/work, and freedom/discipline—the content of the writing course (p. 35). Positioned as a type of cultural analysis that pays particular attention to signifying practices that construct subject positions[5] within systematic constraints such as ideology, economics, and politics, Berlin's social epistemic rhetoric "enables senders and receivers to arrive at a rich formulation of the rhetorical context in any given discourse situation" (p. 84).

Social-epistemic rhetoric affords students the opportunity to revise their subject positions with regard to their own consumption and use of wi-fi technologies by evaluating their own complicity in the cycle of consumerism and creating new positions for themselves as wi-fi users through their writing practices. Social-epistemic rhetoric suggests that teachers confront these issues in the writing classroom and ask students to consider the realities of wi-fi use. To prompt this inquiry, I end with sample assignments that instructors may use to foster critical understandings and uses of wi-fi technologies.

SAMPLE ASSIGNMENTS

1. Have students locate several wi-fi advertisements and analyze them for audience, message, and purpose. Then, ask students to rewrite the ads to make reasonable claims about the availability, accessibility, and reality of wi-fi in their communities.
2. Ask students to review their institution's honor code or code of student conduct and plan for rolling out campus wi-fi hotspots. In groups, have students draft a wi-fi classroom policy that encourages the learning potentials of wi-fi technology.
3. Have students map the wireless coverage in their campus or city using wi-fi sniffers (Kensington makes one for around $20.00), wi-fi laptops, or PDAs. Compare student maps with maps of demographic data—socioeconomic status, median

income, population density—for the same area. Discuss the ideologies of wi-fi deployment and social factors.

4. In a service-learning course, ask students to work with a local community center to establish a "Wireless Community Networks" project, similar to the one in the Chicago area, which is a neighborhood effort to provide low-cost broadband connectivity and related opportunities such as job searching capability and skill development to underserved households, community groups, and small businesses.

5. Assign a semester-long, qualitative research project on student wi-fi use on campus. Options include interviews and observation studies of students to compare student "productivity" levels through wired and wireless access.

6. Have students work through this article and extend the research done here to other consumer electronics products that create particular subject positions for their users: Personal Digital Assistants (PDAs), MP3 players, software packages, video games, mobile telephones, among others.

These assignments will demonstrate the subject positions offered to students in tangible ways and provide them an opportunity to accept, reject, and revise those subject positions from an informed, critical position. More importantly for writing teachers, these assignments reveal that wi-fi can be a critical component in the classroom by engaging students in complex, socially relevant critique and use of wi-fi.

ENDNOTES

1. I am grateful to Ken S. McAllister and Mark Zachry for their generous and insightful comments on this chapter. I am also grateful to Amy C. Kimme Hea and her outside readers for their feedback. Happily, this chapter is another successful publication for the FHE reading group at USU, and I thank them for their support and feedback on this piece.

2. Throughout this chapter, I will assume that readers have a fairly good working knowledge of the technical aspects of wireless networking technologies, at least enough to know that wireless technologies facilitate networking (file sharing, internet access, and other LAN capabilities) through radio signals.

3. Dan Steinbock (2002) glorifies Nokia's employment of globally distributed and shared networks for product research and development across Europe and Asia

(pp. 43-45). He argues that the Nokia model offers new possibilities for flexibility and innovation while maintaining a global market.

4. By "choicing" here I mean the ability of marketers and manufactures working in cooperation to predict consumer desires and appeal to them with reasonable accuracy through market research (discrete choice analysis) and universal product image.

5. Both James Berlin (1996) and I rely heavily on Michel Foucault's development of the concept of the "subject position" which he describes in the *Archeology of Knowledge* (1972) as "a position that may be filled in certain conditions by various individuals" (p. 115). The subject positions that are available to writing students to research and write about are those that are constructed by consumer electronics marketers for consumers and users of wi-fi technologies to fill.

REFERENCES

Alexander, Bryan. (2004, September/October). Going nomadic: Mobile learning in higher education, *EDUCAUSE Review, 39*, 28-35. Retrieved February 2, 2004, from http://www.educause.edu/pub/er/erm04/erm0451.asp?bhcp = 1

Berlin, James A. (1996). *Rhetorics, poetics, and cultures: Refiguring college English studies.* Urbana, IL: NCTE.

Bhave, Mahesh P. (2002, November). Classrooms with wi-fi. *T.H.E. Journal: Technological Horizons in Education.* Retrieved February 2, 2005, from http://www.thejournal.com/magazine/vault/A4218.cfm

Brans, Patrick. (2003). *Mobilize your enterprise: Achieving competitive advantage through wireless technology.* Upper Saddle River, NJ: Prentice Hall.

Chaput, Catherine. (2004, May). *Reality TV and the political economy of fear: Ethical problems and rhetorical solutions.* Paper presented at Biennial Conference of the Rhetoric Society of America, Austin, TX.

Easton, Jaclyn. (2002). *Going wireless: Transform your business with mobile technology.* New York: HarperCollins.

Fisher, Chris. (2002, September 30). New tech key in wireless recovery. *Electronic News, 48,* 16.

Fiske, John. (1996). *Media matters: Race and gender in U.S. politics.* Minneapolis: University of Minnesota Press.

Foucault, Michel. (1972). *The archeology of knowledge* (A.M. Sheridan Smith, Trans.). New York: Pantheon. (Original work published 1971)

Glassner, Barry. (1999). *The culture of fear: Why Americans are afraid of the wrong things.* New York: Basic.

Gohring, Nancy. (2004, February 18). Intel chasing networked home concept. *Wi-Fi networking news: Home entertainment archives.* Retrieved June 10, 2004, from http://wifinetnews.com/archives/002952.html

Haas, Christina, & Neuwirth, Christina M. (1994). Writing the technology that writes us: Research on literacy and the shape of technology. In Cynthia L. Selfe & Susan Hilligoss (Eds.), *Literacy and computers: The complications of teaching and learning with technology* (pp. 319-335). New York: MLA.

Hawisher, Gail E., & Selfe, Cynthia L. (1991). The rhetoric of technology and the electronic writing class. *College Composition and Communication, 42,* 55-65.

Heller, Kevin J. (2003, January 2). WiFi classroom: Boring profs or uninterested students. *Tech law advisor: An Internet and intellectual property policy weblog.* Retrieved February 4, 2005, from http://techlawadvisor.com/2003/01/wifi-classroom-boring-profs-or.html

Hogan, Mike. (2001, May). Why wi-fi? *Entrepreneur Magazine.* Retrieved June 10, 2004, from http://www.entrepreneur.com/article/0,4621,288647,00.html

Intel Centrino. (n.d.) Increase enterprise productivity: Built for wireless. *Notebooks built for wireless.* Retrieved June 9, 2004, from http://www.intel.com/business/bss/products/notebook/centrino/productivity/wireless/index.htm

Janangelo, Joseph. (1991). Technopower and technoppression: Some abuses of power and control in computer-assisted writing environments. *Computers and Composition, 9,* 47-64.

Kitalong, Karla Saari. (2000). "You will": Technology, magic, and the cultural contexts of technical communication. *Journal of Business and Technical Communication, 14,* 289-314.

Lightman, Alex, & Rojas, William. (2002). *Brave new unwired world: The digital big bang and the infinite future.* New York: Wiley.

Lindgren, Mats, Jedbratt, Jörgen, & Svensson, Erika. (2002). *Beyond mobile: People, communications and marketing in a mobilized world.* New York: MacMillan.

Living wireless: How students have unwired themselves. (n.d.). *Living wireless: New mobility offers life beyond the laptop.* Retrieved June 9, 2004, from http://www.intel.com/personal/do_more/wireless/living/student.htm

Mansell, Robin, & Steinmueller, W. Edward. (2000). *Mobilizing the information society: Strategies for growth and opportunity.* New York: Oxford University Press.

Mason, Charles. (1998, September 1). Power to the user: Trends point to core changes in wireless development. *America's Network: Technology for the Information Highway, 102,* 70-73.

Palenchar, Joseph. (2001, November 5). Enhancements to aid CE sales. *TWICE: This Week in Consumer Electronics, 16,* 24.

Parks, Elizabeth. (2004, May 11). Entertainment-centric home pcs and networks to reach $12 billion in sales. *Wi-Fi technology news.* Retrieved June 9, 2004, from http://www.wi-fitechnology.com/displayarticle1154.html

Postman, Neil. (1985). *Amusing ourselves to death: Public discourse the age of show business.* New York: Penguin.

Prensky, Marc. (2005, June/July). What can you learn from a cell phone? Almost anything! *Innovate, 1*(5). Retrieved January 10, 2006, from http://www.innovateonline.info/index.php?view = article&id = 83

Rampton, Sheldon, & Stauber, John. (2003). *Weapons of mass deception: The uses of propaganda in Bush's war on Iraq.* New York: Penguin.

Steinbock, Dan. (2002). Wireless R&D: From domestication to globalization. *Info, 4,* 27-49.

Tangent Corp. (2002). *Wireless classrooms* [Brochure]. Retrieved December 10, 2004, from http://www.tangent.com/products/datasheets/MobileClass.pdf

United States Senate Committee on the Judiciary. (2001). *The AOL/Time Warner Merger: Competition and consumer choice in broadband internet services and technologies* (DOCID: J-106-66). Washington, DC: U.S. Government Printing Office.

Wireless Community Networks. (n.d.). A project of the center for neighborhood technology. Retrieved December 10, 2004, from http://wcn.cnt.org/

5

RETERRITORIALIZED FLOWS

CRITICALLY CONSIDERING STUDENT AGENCY
IN WIRELESS PEDAGOGIES

Melinda Turnley

We computer compositionists are accustomed to hearing exaggerated promises for new technologies to democratize education and bring literacy to marginalized groups. Recently, wireless and mobile technologies have been forwarded as a means to revolutionize learning experiences within classrooms, across campuses, and beyond traditional academic spaces. Much like early idealizations of distance education and other web-based pedagogies, discussions of mobile learning often highlight possibilities for increased student agency and access. Wireless educational practices seemingly deterritorialize classroom spaces by transforming how, when, and where learning can occur. Thus far, conversations in the field have highlighted the role of wireless technologies within classroom environments. For instance, scholars have explored how wireless classrooms impact our negotiations of physical and social spaces (Meeks, 2004) and reconfigure the ways we conceive of and occupy locations on campus (Zoetewey, 2004). As we consider the potentials of wireless pedagogies, however, we should examine educational practices not only within classrooms but also those outside traditional academic spaces. Many colleges and universities highlight distance education as part of their instructional initiatives, and wireless technologies seem like a fitting addition to these ventures. Although wireless technologies could contribute to off-campus learning, we should remain mindful that they do not automatically enhance access or mobility.

On the surface, wireless technologies seem like an ideal solution for nontraditional and underserved students. Creating wireless networks often is more cost effective than laying additional land lines for phone and Internet access. Therefore, proponents often present wireless technologies as offering increased information access for rural and economically challenged areas. For example, in October 2005, Microsoft announced a RFP in support of "deep academic research" on issues of technology access for underserved populations (Microsoft Research, 2005 ¶ 11). This "Digital Inclusion Through Mobile and Wireless Technologies Research Funding Initiative" explicitly forwards wireless and mobile technologies as powerful means to extend traditional user bases and expand "the capabilities of computing technology worldwide to better serve social and economic challenges of underserved communities, both rural and urban" (Microsoft Research, 2005, ¶ 1). Similarly, mobile learning is framed as an ideal option for students who must work full time or who are unable to leave their local communities to attend school. Forwarding such potentials, companies such as IBM, Cisco, and Toshiba tout wireless networks as fostering ubiquitous learning through anytime, anywhere access. They market wireless technologies as customizable solutions that deliver mobility, convenience, efficiency, productivity, flexibility, and ease of use. Toshiba's web site (n.d.) on mobile computing in higher education, for example, declares that "education is everywhere" and sets forth the goal to "transform virtually any space into a fully functional learning environment" (Toshiba, ¶ 3).

Visions of education that are less place bound certainly have pedagogical potential worth exploring. Utopic framings of limitless "anytime, anywhere" education, however, problematically assume that technology determines space—that technical access equates with all types of access. This approach deterministically posits that wireless technologies can render space controllable and transparent. Removing wires from our machines, however, does not erase their ties to institutional and political concerns. Even if wireless networks are less costly to implement than wired ones, they still are neither free nor self-sustaining; mobile technologies are not quick fixes for problems of access. We composition teachers and program administrators, therefore, should be careful about the roles that we envision for students as we consider wireless possibilities. Deterministic frameworks present wireless networks as neutral spaces that automatically equalize all environments and student positions. Such approaches present learning as an individualized, decontextualized process and thus serve to reinscribe social hierarchies surrounding issues of access and academic success.

To counter such determinisms, we need to incorporate issues of space into our critical discussions of technology and highlight the construction of space as a social, material process. Our considerations of space should examine the impact of mobility on issues of access. Wireless technologies have the potential to transform educational goals, relationships, and practices in productive ways, but they also can be used to reinscribe traditional hierarchies and exclusions. Rather than embracing narratives of automatic efficiency and autonomy, I suggest that our wireless pedagogies frame student agency as intersubjective and encourage multiple perspectives on shared contexts. In their critique of psychoanalytic and Marxist conceptualizations of subjectivity and desire, Gilles Deleuze and Félix Guattari offer a theory of postmodern space that emphasizes local connections as the crucial element of productive mobility. They posit that, rather than offering culmination points or external ends, structures should maintain open, destratified, rhizomatic, middle spaces (1987, pp. 478-479). They assert that traditionally hierarchical spaces should be deterritorialized so they allow for multiple trajectories and unobstructed movement. Seemingly, this vision supports idealizations of wireless technology. Deleuze and Guattari, however, also theorize hindrances to nomadic "total flow." Deterritorialization disrupts established assumptions and structures and thus offers possibilities for new forms of agency. This productive force, however, works in tandem with the counter-tendency of reterritorialization, or the potential to be assimilated back into dominant frameworks (1987, p. 54). Deleuze and Guattari suggest that, even as selves pursue "lines of flight" away from hegemonic structures, they can be reabsorbed. In particular, if desires are framed as autonomous and universal rather than relational and local, the potential for multiplicity and new forms of agency is blocked.

In the discussion that follows, I consider risks of reterritorialization in our framings of mobility and wireless education. Countering the deterministic premise that mobile technologies are automatically inclusive, I maintain that we should attend to wireless networks' imbrication with cultural networks. We should be conscientious in our positioning of students so that possibilities for remote, mobile participation are not channeled back into structures of social and economic exclusion. To this end, I begin by engaging computers and composition scholarship on the roles of students in the diffused environments of distance learning. As these scholars argue, computer-mediated changes in pedagogical time and space have the capacity to further exclude already marginalized student populations. I then turn to conversations that examine connections among technology, access, and literacy. The interrelated axes of socioeconomic status, race, and education significantly influence the distribution of technological resources and training. Thus, issues of access are deeply tied to larger cultural and material contexts and include more than just questions of hardware and software. To situate my consideration of access issues, I reflect on my experiences teaching at New Mexico State University (NMSU), a land grant institution in a relatively poor county in one of the most economically challenged states in the country. This case helps to illustrate how transformations of educational spaces involve more than just technical factors. As a way to critically intervene in these processes, I turn to Deleuze and Guattari's concepts of de- and reterritorialization. Their theorization of productive mobility emphasizes the importance of connection, collectivity, multiplicity, and localization rather than universalization. By engaging dominant narratives and resistant practices in explicitly spatial terms, they offer nondeterministic means for refiguring student agency in relation to wireless composition pedagogies. I conclude by drawing upon this framework to raise issues significant for our construction of deterritorialized—situated, collaborative, ethical—roles for students within wireless pedagogies.

DISTANCE EDUCATION

Scholars already have begun to consider spatial refigurations of writing pedagogies in their research on distance education. As Ann Hill Duin (1998) suggests in her discussion of technical communication faculty, distance education should be thought of as a cultural change that influences our courses, programs, institutions, and educational mission. Offering fully or quasi-distance courses affects not only the teaching done in specific sections but also the curricula, allocation of resources, and overall agendas of writing programs. Researchers have discussed the ways in which synchronous and asynchronous online learning environments shift our communication practices (Ragan & White, 2001), course design (Blythe, 2001; Savenye, Olina, & Niemczyk, 2001; Thrush & Young, 1999), and models for funding (Werry, 2002). Additionally, scholarship has addressed the impact of distance education on student and teacher identities (Miller, 2001), teacher presence and workload (Hailey, Grant-Davie, & Hult, 2001), the roles of untenured faculty (Samuels, 2004), and the demographics of student populations (Schneider & Germann, 1999).

Such research reinforces the notion that extending instruction beyond the traditional boundaries of classroom spaces raises a range of practical and conceptual concerns. Technologies influence and are influenced by larger institutional, cultural contexts. Thus, moving courses either partially or fully online involves not only technological but also social changes. Further, although distance education raises new questions for us to consider, it does not completely remake the educational landscape. Certain ideological and ethical concerns persist in relation to our assumptions and methods. As Kristine Blair and Elizabeth Monske (2003) assert, rather than rushing into curricular integration, we should carefully consider who benefits, or does not, from the creation of distance learning environments (p. 442). Benefits associated with distance education can include enhanced learning and reductions in instructional cost, yet these goals often run counter to one another (Brady, 2001). Online courses are marketed as a means to accommodate a diverse population of learners and make higher education more accessible to nontraditional students (Webb Peterson, 2001). Distance education, however, also has the potential to reproduce the very exclusions that it is intended to remedy.

The seeming diffusion and disembodiment of online instruction can limit opportunities for collaboration among class participants. Thus, to succeed in distance courses, students generally must be self-directed, comfortable, and confident in their work (Miller, 2001). The need for this sort of autonomy can create challenges for any student, but students with limited technology resources and training are likely to face additional complications in regard to their participation (Webb Peterson, 2001). A wider range of students may be able to enroll in distance courses, but initial matriculation does not insure completion. "Education has always created socioeconomic distances by reproducing the existing class structure even as it has held out the hope of social mobility; distance education recreates different kinds of material and social access chasms even as it promises to bridge gaps of time and space" (Brady, 2001, p. 356). Many discussions of wireless learning echo the promises of equity and access prominent in pushes for increased distance education. As noted, companies such as Toshiba, Cisco, and IBM assert that the mere presence of wireless technologies can turn any space into an educational environment. Such assumptions extend problematic claims about the power of distance education to encompass all spaces at all times. PDAs, laptops, cellular phones, and other wireless devices, however, do not guarantee increased access for a broader range of students. As we incorporate mobile technologies into our composition curricula, we should remain mindful of how social hierarchies influence student success. Previous research on distance education provides productive points of reference for this sort of critical consideration.

ISSUES OF ACCESS

Discussions of access also offer valuable insight into the complex relationships among technology, socioeconomic factors, and education. Access to emerging technologies is a function of cultural factors such as income, education, race, and social class (Grabill, 1998, 2003; Moran, 1999; Selfe, 1999). Computer resources are distributed differently among different social groups, and "the technopoor's relation to institutional access with computer technologies is one of hierarchical exclusion" (Grabill, 2003, p. 460). Such exclusion, of course, relates to equipment, but this factor is only one aspect of access. Access involves not only infrastructural resources but also literacy and social acceptance (Grabill, 1998; Porter, 1998). In our culture, both technology and literacy are entwined with narratives of progress and equality; they are praised as catalysts for improved education and increased social mobility and economic prosperity. Pursuing more egalitarian educational practices is not simply a matter of access to equipment or acquisition of narrowly defined sets of literacy skills. These sorts of instrumentalist strategies do not adequately address larger patterns of exclusion. Therefore, we must pay critical attention and consider the ways in which responsible approaches to technology encompass more than just issues of use (Haraway, 1991; Selfe, 1999).

Charles Moran (1999) aptly addresses the need for computer compositionists to consider the material and cultural contexts of teachers' and students' work. The gap between technological haves and have-nots is widening, and thus we should make an ethical commitment to intervene critically in issues of access. Dominant assumptions about the power of new technologies often highlight issues such as ease, efficiency, transparency, and prosperity. Rather than accepting deterministic framings that universalize technological benefits, we should situate our pedagogical decisions within both local and larger cultural contexts. Drawing upon Donna Haraway's notion of situated knowledges, Cynthia L. Selfe (1999) argues that we must look for particular sites in which we can practice informed action. In order to address the complex linkages among technology, literacy, poverty, and race, we need to allow for a wide range of positionings and find ways to "to construct a 'larger vision' of our responsibilities as a profession, one that depends on a strong sense of many somewheres (e.g., schools, classrooms, districts, communities)" (p. 429). This emphasis on situatedness seems especially relevant to discussions of mobile learning. Wireless technologies may allow individuals to study in locations outside of traditional campus environments, but students' positionings involve more than just physical space. Students also occupy larger institutional and cultural spaces, ones that are not automatically transformed by technology. Therefore, instead of assuming that social mobility inherently accompanies technical mobility, our development of wireless pedagogies should explore approaches to space which account for both blockages to and openings for agency.

MY SENSE OF SOMEWHERE

Heeding this call for situatedness in our work with technology, I reflect on my recent experiences teaching at NMSU. I lived and worked for three years in New Mexico, a state where issues of income and access are prominent, undeniable concerns. The Children's Partnership, a national nonprofit, nonpartisan organization, has compiled statistics from the U.S. Departments of Labor and Commerce, the Bureau of Labor Statistics, and the U.S. Census Bureau into a fact sheet about the status of youth and technology in New Mexico (Children's Partnership). According to this list of figures, New Mexico ranks 48th among the 50 states at combating child poverty. Forty-nine percent of households in the state do not own a computer, and 57% do not have Internet access; both of these percentages are below national averages, which are 43% and 49%, respectively. In fact, 18% of New Mexican children do not even have a phone at home. An increasing number of schools are providing at least some access to computers and the Internet. Yet, teachers do not always receive adequate training to support student learning in networked environments. In 33% of schools in New Mexico, the majority of teachers (at least half) identify as "beginners" when it comes to using technology; this is well above the national average of 24% (Children's Partnership).

Being on the Rhetoric and Professional Communication faculty in the English Department at NMSU, I had to remain sensitive to the fact that many students at this institution have limited technological experience and resources. NMSU currently is the only Carnegie I land-grant institution that is also designated a Hispanic-Serving Institution. Additionally, many NMSU students come from low income households in rural areas and are the first generations of their families to attend college. Based on these demographics, this student population is at a high risk to fall through the digital divide.

According to the National Telecommunications and Information Administration's 1999 online report, "Falling through the Net: Defining the Digital Divide," Internet access gaps exist between white and Hispanic households, urban and rural areas, and high-and low-income levels, and high-and low-education levels (National, 1999, Executive Summary, ¶ 4). Further, the administration's follow up report in 2000 indicates that, although the overall level of U.S. digital inclusion is rapidly increasing, the gap continues to widen for blacks and Hispanics (National, 2000, Executive Summary, ¶ 3).

Some students at NMSU, of course, are technically savvy and have access to a range of technologies in a range of locations. Many others, however, are relative novices who never have had consistent access to computers and the Internet. Even in my upper-division writing courses, class members often self-identified as lacking proficiency in seemingly "basic" technologies such as word processing, e-mail, and the WWW. For example, I had a student drop one of my 200-level technical communication courses because it met regularly in a networked classroom and required students to send and receive email attachments. I assured him that I would support his participation and help him learn unfamiliar course technologies, yet he still opted to drop and re-enroll in a noncomputerized section. This lack of technological confidence was common in the courses I taught at NMSU. Each semester, some of my students were anxious about computer-mediated learning, even within the collaborative context of a classroom in which they were supported by a teacher and other students.

IBM promotes the potential for "pervasive computing" to expand an institution's ability to enhance revenues while offering more courses to nontraditional students (IBM, ¶ 1). Yet, such assumptions are dangerous because they elide connections between student backgrounds and success. Further, in their marketing of wireless educational solutions, Cisco asserts that students now are totally dependent on access to technology, and pervasive technology is part of "basic student survival" (Cisco, ¶ 3). This framing, however, is not commensurate with my sense of students at NMSU. As of spring 2006, my last semester on campus, the wireless coverage was limited to select buildings, and students typically did not own devices that allowed them to use the network. Further, many students did not even have consistent off-campus access to wired networks; for them computer technology was far from pervasive. In relation to mobile learning, we computer compositionists should question claims, like Toshiba's, that "education is everywhere." We must remain mindful that, for some students, even traditional academic spaces pose educational challenges.

Like many other campuses, NMSU struggles with relatively low retention rates. Because of the demographics of this student population, we faced additional challenges in relation to attrition. Thus, reflecting on my participation in this local situation, I am concerned about the roles of students in wireless pedagogies. On the one hand, people in rural areas might be better able to enroll in university courses if they had wireless access to online courses. Yet, how productive would these educational experiences be if new students did not receive adequate support and training? Students on campus with face-to-face support often have trouble staying enrolled and successfully completing degrees at NMSU. I question how well students would fare left on their own to individually negotiate technical and educational challenges.

I would not suggest that institutions such as NMSU should reject wireless or mobile learning. Likewise, I do not mean to imply that all institutions are like NMSU or that NMSU is singular. Instead, like Cynthia L. Selfe and others, I argue that we must be careful in our design and implementation of new learning environments. We should be mindful of correlations among technology, socioeconomic factors, and access and take seriously the ways in which they play out in our particular "somewheres." Further, we should be critical of ways in which determinisms concerning the ease and efficiency of wireless technology could exacerbate exclusionary educational structures. We need ways to theorize mobility that support, rather than isolate, students. As one potential approach, I offer the work of Deleuze and Guattari. Many productive approaches are available that frame space as a social practice that involves multiple perspectives. I turn to this particular framing because Deleuze and Guattari explicitly define agency in relation to movement and highlight the figure of the nomad; they discuss not only the potential for the dynamic restructuring of space but also the possibility for affirmative, relational movement.

DELEUZE AND GUATTARI'S RETERRITORIALIZATION AND DETERRITORIALIZATION

Spaces and our relationships to them are not neutral. Wireless technologies have the potential to reshape the geography of education, but such reconfigurations are not automatically libratory. As we have learned from distance education, moving instruction outside the shared time and space of traditional classrooms can isolate students and reinforce social disparities.

We computer compositionists thus need to explore strategies for critically engaging mobile learning and complicating our sense of educational space. The writings of Deleuze and Guattari offer a rich approach to theorizing mobility that accounts for both the furtherance of and blockages to agency. Their framing is particularly relevant to discussions of wireless technologies because it highlights spatial figures. To explore alternative social configurations, they deploy a range of space-related metaphors, including the tropes of territories and nomads. Deleuze and Guattari collaboratively authored *Anti-Oedipus* (1972) and *A Thousand Plateaus* (1987). These two volumes constitute *Capitalism and Schizophrenia*, their critique of psychoanalytic and Marxist approaches, which they see as forwarding repressive power structures and symbolic apparatuses. Their goal is to disrupt the binary logic, regimented individualism, negativity/lack, and universalization associated with repressive bureaucracies of the traditional family, state, and nation.

Because they want to critique traditional organizational structures, Deleuze and Guattari need an alternative, more flexible term for describing sets of social relations. They use the spatial notion of a territory to designate relationships that are formed by the interactions of a group. Territorialized spaces are not simply physical locations but sets of relations that encourage and discourage certain forms of agency and desire. Territories are calcified sets of practices that organize centrally around rubrics such as art, education, religion, and the state. Forces generated by lived experience are diverse, but territories channel these energies narrowly so that we experience them more homogenously. Because territories are constructed rather than essential, their totalizing structures can be disrupted, or deterritorialized. Such ruptures can open up productive "lines of flight," but they are not inherently library or absolute (1987, p. 55). Deleuze and Guattari also introduce possibilities for reterritorialization. Deterritorialization is an open-ended process that encourages new connections and variations, whereas reterritorialization fixes movement along lines of hierarchy and centralization (1987, pp. 508-510). Reterritorialization channels potential back into familiar territories; it is the resurrection of old meanings and the reinscription of relations based on exclusion and segregation. Deleuze and Guattari's associational, postmodern style does not render concise definitions for terms such as de- and reterritorialization. Instead, meanings accumulate through the authors' iterative explorations of various concepts throughout their books. To help clarify my use of these terms, I have tracked references across different moments in their works and synthesized them into Table 5.1.

TABLE 5.1. FORCES OF DETERRITORIALIZATION AND RETERRITORIALIZATION

TERM	TENDENCIES	EXAMPLE
Territory (Relationships formed by interactions of a group)	Exclusion, universalization, reduction. Organizes diverse forces so they are channeled into seeming unities.	Educational practices that define literacy and success in relation to individual student initiative without considering larger socioeconomic patterns of exclusion.
Deterritorialization (Open-ended process that destabilizes territories)	Disruption, reorientation, multiplication, localization. Encourages new connections, changes in direction, and potentials for agency.	Use of new resources (such as mobile technologies) to disrupt old hierarchies and facilitate flexible, dynamic educational opportunities for more diverse groups of students.
Reterritorialization (Resurrection of old meanings and structures to shore up ruptures)	Cooptation, appropriation, reinscription. Blocks agency through reassertion of relations based on segregation, hierarchy, and centralization.	Potential for change thwarted by redeployment of old frameworks such as technological determinism and individualistic, exclusionary approaches to access and literacy.

Deleuze and Guattari correlate both of these processes with certain types of environments. Striated space is instituted by the power structures of state apparatus and promotes reterritorialization through separation, totalization, and hierarchy. Smooth space, however, has a greater power for deterritorialization because it allows for uninterrupted movement. It permits localized changes in multiple directions and thus fosters limitless possibilities for making new connections. These spaces are different in nature but not simply opposed or mutually exclusive; they are interrelated and each is constantly being translated into the other. "Of course, smooth spaces are not in themselves libratory. But the struggle is changed or displaced in them, and life reconstitutes its stakes, confronts new obstacles, invents new paces, switches adversaries. Never believe that a smooth space will suffice to save us" (Deleuze & Guattari, 1987, p. 500). Any movement potentially can disrupt a territory and create a smooth space, but such openings for change are temporary and contingent. If positions become fixed, they block access and agency. Thus, Deleuze and Guattari need a new sense of the self that reflects their commitment to mobility and change. They introduce the figure of the nomad—a self whose trajectory cuts across, and potentially disrupts, multiple territories. Unlike the linear, predetermined path of oedipal self, the path of the nomad resists totalization and stagnation, follows various paths, and is constituted by diverse connections. "The life of the nomad is in the intermezzo" (Deleuze & Guattari, 1987, p. 380).

For Deleuze and Guattari, agency is an intersubjective process that develops through multiple, flexible relationships. As an alternative to controlling, hierarchical systems, they propose an antifoundational approach that forwards an affirmative, relational sense of self. Psychoanalysis, Marxism, and other totalizing approaches envision groups as organic bonding among hierarchized individuals. Such theories define the self in relation to individual consumption, or the filling of a lack. Deleuze and Guattari, conversely, are interested in the positive process of production. For them, desire is not purely individual; they view both selves and groups as collective. Their approach values multiplicity rather than unity, continuous variation rather than static homogeneity, trajectories rather than points, events rather than things, performance rather than competence, and rhizomatic rather than arboreal structures. They seek flexible consolidations, "*becomings*, which have neither culmination nor subject, but draw one another into zones of proximity or undecidability" (Deleuze & Guattari, 1987, p. 507).

Productive potentials for deterritorialization must be consciously and continuously cultivated. Lines of flight away from familiar educational territories can be facilitated by new technologies, but these potentials can also be co-opted. Therefore, we need to complicate the assumption that mobility, like that made possible through wireless technologies, renders space neutral or universal. In fact, nomadic movement is productive because it does just the opposite; it multiplies our relations to space and each other, rather than reducing them. Deleuze and Guattari (1987) argue that the development of smooth spaces is "achieved not in a centered, oriented globalization or universalization but in an infinite succession of local operations" (p. 383). That is, actions should be situated in relation to specific human relations, in specific locations, at specific times. Nomadic space is localized without being rigidly delimited. Even within shifting landscapes, mobile, multiple selves can consolidate and converge. As we disrupt older educational territories through the incorporation of wireless technologies, we need to strategize ways to foster new groups and connections rather than individualizing agency and desire. Hierarchies are not automatically flattened by technological respatializations. Thus, removing students from traditional campus-based communities without resituating them within new ones can reinscribe exclusions and reterritorialize the potentials for mobile learning.

IMPLICATIONS FOR MOBILE PEDAGOGIES

As I have suggested, companies often market wireless technologies as turning any space into a universal site for learning. Such approaches are troubling because they totalize rather than localize technology use and thus elide cultural and institutional factors related to education, access, and literacy. Mobile learning does have the potential to destabilize traditional academic territories and provide more flexible, inclusive opportunities for a wider range of students. Nevertheless, in addition to deterritorialization, reterritorialization is also a possible result of shifts in our pedagogical spaces. Wireless technologies may change the terrain of our classrooms, but they do not guarantee nomad agency. In fact, such changes are likely to reinforce old norms if they are made uncritically and deterministically. As we consider pedagogical applications for laptops, PDAs, cell phones and other mobile technologies, we need to make certain to frame our efforts in terms of production rather than just consumption. Otherwise, we run the risk of limiting student agency and success. We should be wary of "anytime, anywhere" frameworks that characterize students as consumers of reified, individualized products. Such positionings echo the neurotic, oedipal self based on negative lack critiqued by Deleuze and Guattari. Instead, we should position students as participating in relational events and producing new connections and relationships. As mobility allows for multiple teaching sites and processes, we should conceive of new ways to encourage collective action and collaborative experiences. That is, we should seek to create smooth, deterritorialized spaces in which students can engage in learning as nomad selves.

The theories of Deleuze and Guattari are complex in their articulation of postmodern space as multiple, continuously variable sets of connections. Therefore, my account of their approach cannot be comprehensive. I, however, hope that I have given a constructive sense of their critique of dominant structures and the roles of selves in relation to them. Further, I hope that my overview begins to suggest ways in which their nomadology could frame critical approaches to wireless technologies. As Cynthia L. Selfe suggests (1999), we need to locate our work with technology in relation to particular sites of practice. Wireless and mobile technologies complicate our "sense of somewhere" by disrupting traditional territories related to educational space. In light of such disruption, we must be careful not to decontextualize our pedagogies in ways that reinforce the digital divide. We need to complicate determinisms that seem to universalize access to technology and literacy.

In order to resist this sort of exclusionary universalization, we should continue to situate our teaching in relation to local rhetorical and institutional situations. Even if students are participating in courses from various locations, we still can encourage a sense of common purpose and shared context. We must also maintain a complex sense of access that accounts for not only students' individual acquisition of equipment but also larger cultural issues related to technological and academic literacies. The design and implementation of wireless pedagogies needs to address economic issues related to technology resources as well as social issues related to student preparation and retention. In the state of New Mexico, a significant number of children and adults live in rural areas or on reservations. Providing traditional network access to these populations often is costly, and thus they likely could benefit from wireless resources. Simply creating Wi-Fi access points and supplying machines, however, would not address adequately the range of other issues related to attrition rates among these groups at NMSU and other universities. No technology is transparent or automatically democratizing. Therefore, we need to think broadly in terms of our support for literacy and technology. Giving students technical access to college courses from multiple locations does not necessarily improve their chances of successfully completing a degree, or even a single course. In fact, proclaiming ease and equity for students without also constructing a sustainable plan for their participation could actually serve to further marginalize them.

FOSTERING LOCAL CONNECTIONS
AND INTERSUBJECTIVE AGENCY

In conclusion, I offer some issues to consider as we explore pedagogical applications for wireless technologies. Although specific factors vary from institution to institution, questions of student positioning remain a constant concern. Online environments already have begun to complicate our sense of educational space, and wireless devices have the potential to extend these dynamics even further. Rather than uncritically embracing the promise of increased student mobility and agency, we should reflect on our assumptions about the role of space within writing instruction. We need to remain mindful that our constructions of space—both physical and virtual, with and without wires—encourage certain relations while inhibiting others. Mobile learning offers increased possibilities for variation, multiplicity, and flexibility in our teaching. It, however, should not reduce literacy instruction to idiosyncratic, completely individualized methods. Even if students' learning spaces can become more customizable, their experiences still are situated within larger cultural frameworks for education and literacy. Education and communication are relational, contextualized processes, and we should continue to highlight this situatedness as we respatialize instruction. As we set up wireless pedagogies, we should emphasize the importance of connection in both the means and ends of our courses. The notion of anytime, anywhere instruction implies an individualized model in which students can move through discrete activities at their own paces. Though this approach might seem convenient, it runs counter to our field's sense of writing and learning as relational, rhetorically situated activities. Like the nomadic movement described by Deleuze and Guattari, student agency should be framed as intersubjective. Our pedagogies should encourage smooth spaces which allow for the convergence of multiple paths rather than striated spaces that isolate students along autonomous, unilinear tracks. Exploring potentials for wireless technologies, we need to approach mobility so that it is not co-opted back into educational territories which instrumentalize technology and literacy.

Dominant narratives about wireless technologies proclaim that education is everywhere. Yet, informed action must be grounded in relation to particular sets of social arrangements. We need to help students construct a sense of the local, even if they are not sharing the same physical space or synchronizing all of their actions. As discussions of distance education have suggested, students in disembodied or dispersed learning environments easily can feel isolated. Students already at high risk for dropping out, in particular, can become discouraged and fall behind. Thus, to help students develop a sense of agency, we must find techniques for helping them feel connected to courses and each other in multiple ways. At NMSU, for example, many students face multiple blocks to successful participation in college-learning environments, computer-mediated or not. As I indicated previously, lack of technical confidence was a significant obstacle for numerous students in my writing courses at this university. This blockage to agency and access cannot be addressed simply by giving students laptops, PDAs, or other wireless devices. Students' successful participation in literate practices requires that they be invested members of learning environments. Thus, as we expand wireless opportunities both on and off campus, we must help students develop critical senses of themselves as positioned within not only technological networks, but complex social ones as well.

One means for accomplishing this goal is to make certain that a course has common points of reference. In addition to common materials, students should have reciprocal methods for communicating with teachers and other students. In our course descriptions, policies, projects, requirements, and instructional activities, we should value productive interactions among members of our classes. Students' remote and/or mobile participation will, of course, pose challenges for collaboration. Yet, encouraging investment in this sort of work is important to not only students' development as communicators but also their overall success in school and future professional lives. Drawing upon Deleuze and Guattari, we should frame our teaching spaces around events rather than just things. Students certainly will continue to compose products for evaluation, but this should not be their only productive role. They should also see themselves as following multiple, convergent trajectories. Instead of simply moving from one point in the curriculum to another, they should be encouraged to explore dynamic connections between their work and the work of other students in the class. Whether it's through the use of text messages on cell phones, wireless Internet access to blogs, or other means, we need to continue to create a sense of shared context for students.

To foster these sorts of diverse combinations, courses could include cohort groups or partnerships that encourage students to seeing learning and communication as relational activities. These sorts of networks could provide students with both flexibility and support. Students could belong to multiple groups, and members of each collective could collaboratively set agendas and schedules. Team-based learning could also facilitate critical reflection on issues of space and positionality. Course activities could ask students to connect work in the course to experiences in other virtual and physical contexts. The local situation of the class could be expanded to incorporate the various places from which students are participating. Technical and spatial issues easily can remain seemingly invisible, yet highly influential, aspects of a class.

Countering this deterministic tendency could have both practical and intellectual benefits. Explicit discussions about spatial concerns would provide students with opportunities to voice concerns or problems and receive support from both peers and the instructor. Such conversations could also help students complicate their understandings of these concerns and thus further develop their technological and spatial literacies. As our machines become smaller, lighter, and more ubiquitous, they can also become more difficult to see politically and materially (Haraway, 1991, p. 153). Rather than letting wireless technologies function ethereally and invisibly, we should highlight their roles in the content of our courses. Projects could ask students to critically engage their local spaces; assignments could prompt students to literally and figuratively locate themselves as members of the various overlapping communities that converge among members of the class. In fact, issues of space could provide a productive, overarching theme for a rhetoric class. Students could critically consider the ways in which textual and information design is spatialized, or how cultural and institutional spaces impact communication.

Issues of access and literacy are enmeshed with our cultural and material contexts. Hype surrounding new technologies often promises simplified learning and teaching, but we computer compositionists know that such claims are reductive and dangerous. As we develop wireless pedagogies, we need to focus critical attention on issues of space and avoid problematic reterritorializations of older, exclusionary hierarchies. Engaging theories such as Deleuze and Guattari's is a productive starting point for exploring strategies that maintain our and students' senses of multiple somewheres. Such care can help us develop opportunities for deterritorialized mobile learning and nomadic student agency.

ACKNOWLEDGEMENTS

I wish to extend special thanks to my former students and colleagues at New Mexico State who made my experiences there so professionally and personally rewarding. I am also grateful to Amy Kimme Hea for her insightful comments and support of this chapter.

REFERENCES

Blair, Kristine L., & Monske, Elizabeth A. (2003). Cui bono?: Revisiting the promises and perils of online learning. *Computers and Composition, 20*, 441-453.

Blythe, Stuart. (2001). Designing online courses: User-centered practices. *Computers and Composition, 18*, 329-346.

Brady, Laura. (2001). Fault lines in the terrain of distance education. *Computers and Composition, 18*, 347-358.

Children's Partnership. (n.d.). New Mexico youth and technology fact sheet. Retrieved January 11, 2005, from http://www.childrenspartnership.org/young americans/statefacts_nm.html.

Cisco Systems. (n.d.). About cutting the cord: Enabling anytime, anywhere access. Retrieved January 12, 2005, from http://www.cisco.com/en/US/strategy/education/higher_ wireless.html

Deleuze, Gilles, & Félix Guattari. (1972). *Anti-Oedipus: Capitalism and schizophrenia* (Robert Hurley, Mark Seem, & Helen R. Lane, Trans.). New York: The Viking Press. (Original work published 1972)

Deleuze, Gilles, & Felix Guattari. (1987). *A thousand plateaus: Capitalism and schizophrenia* (Brian Massumi, Trans.). Minneapolis: University of Minnesota Press. (Original work published 1980)

Grabill, Jeffrey T. (1998). Utopic visions, the technopoor, and public access: Writing technologies in a community literacy program. *Computers and Composition, 15,* 297-315.

Grabill, Jeffrey T. (2003). On divides and interfaces: Access, class, and computers. *Computers and Composition, 20,* 455-472.

Hailey, David E., Grant-Davie, Keith, & Hult, Christine A. (2001). Online education horror stories worthy of Halloween: A short list of problems and solutions in online instruction. *Computers and Composition, 18,* 387-397.

Haraway, Donna J. (1991). A cyborg manifesto: Science, technology, and socialist-feminism in the late twentieth century. In *Simians, cyborgs and women: The reinvention of nature* (pp. 149-181). New York: Routledge.

Hill Duin, Ann. (1998). The culture of distance education: Implementing an online graduate level course in audience analysis. *Technical Communication Quarterly, 7,* 365-388.

IBM. (n.d.) IBM wireless infrastructure solution for higher education. Retrieved January 11, 2005, from http://www-1.ibm.com/industries/education/doc/content/solution/ 366023310.html

Meeks, Melissa Graham. (2004). Wireless laptop classrooms: Sketching social and material spaces. *Kairos, 9*(1). Retrieved January 4, 2006, from http://english.ttu.edu/kairos/9.1/binder2.html?coverweb/meeks/index.html

Microsoft Research. (2005, October). Digital inclusion through mobile and wireless technologies research funding initiative. Retrieved January 4, 2006, from http://research.microsoft.com/ur/us/fundingopps/RFPs/DigitalInclusion_2005_RFP.aspx

Miller, Susan. (2001). How near and yet how far? Theorizing distance teaching. *Computers and Composition, 18,* 321-328.

Moran, Charles. (1999). Access—The "A" word in technology studies. In Gail E. Hawisher & Cynthia L. Selfe (Eds.), *Passions, pedagogies, and 21st century technologies* (pp. 205-220). Logan: Utah State University Press.

National Telecommunications and Information Administration. (1999). Falling through the net: Defining the digital divide. Retrieved February 2, 2005, from http://www.ntia.doc.gov/ntiahome/fttn99/contents.html

National Telecommunications and Information Administration. (2000). Falling through the net: Toward digital inclusion. Retrieved February 2, 2005, from http://www.ntia.doc.gov/ntiahome/fttn00/contents00.html

Porter. James E. (1998). *Rhetorical ethics and internetworked writing.* Greenwich, CT: Ablex.

Ragan, Tillman J., & White, Patricia R. (2001). What we have here is a failure to communicate: The criticality of writing in online instruction. *Computers and Composition, 18,* 399-409.

Samuels, Robert. (2004). The future threat to computers and composition: Nontenured instructors, intellectual property, and distance education. *Computers and Composition, 21,* 63-71.

Savenye, Wilhelmina, Olina, C. Zane, & Niemczyk, Mary. (2001). So you are going to be an online writing instructor: Issues in designing, developing, and delivering an online course. *Computers and Composition, 18,* 371-385.

Schneider, Suzanne P., & Germann, Clark G. (1999). Technical communication on the web: A profile of learners and learning environments. *Technical Communication Quarterly, 8,* 37-48.

Selfe, Cynthia L. (1999). Technology and literacy: A story about the perils of not paying attention. *College Composition and Communication, 50,* 411-436.

Thrush, Emily A., & Young, Necie Elizabeth. (1999). Hither, thither, and yon: Process in putting courses on the web. *Technical Communication Quarterly, 8,* 49-59.

Toshiba. (n.d.). Higher education. Retrieved January 12, 2005, from http://www.toshiba.ca/web/link?id = 1100

Webb Peterson, Patricia. (2001). The debate about online learning: Key issues for writing teachers. *Computers and Composition, 18,* 359-370.

Werry, Chris. (2002). The work of education in the age of ecollege. *Computers and Composition, 19,* 127-149.

Zoetewey, Meredith. (2004). Disrupting the computer lab(oratory): Names, metaphors, and the wireless writing classroom. *Kairos, 9*(1). Retrieved January 4, 2006, from http://english.ttu.edu/kairos/9.1/binder2.html?coverweb/zoetewey/index.html

PART III

Cutting the Cord: Stories on Wireless Teaching & Learning In The Composition Classroom

6

FROM DESKTOP TO LAPTOP

MAKING TRANSITIONS TO WIRELESS LEARNING
IN WRITING CLASSROOMS

Will Hochman
Mike Palmquist

For writers and writing instructors, innovative writing tools are cause for excitement and concern. Excitement arises from the hope of new solutions to old problems (such as replacing white-out as the writer's primary revision tool) and anticipation of new ways to ease the life of the writer (such as WYSIWYG editors, commenting tools, and hyperlinks, among many other innovations). Concern emerges from the dawning realization that, no matter how well we have assessed and, as appropriate, incorporated the last round of innovations into our writing and teaching, there is even more to learn.

The latest wave of innovation has brought the widespread availability of portable, networked computing devices. Freed from the short tether of telephone lines and LAN cables, writers are composing, researching, and sharing their work through wireless laptops, handhelds, and tablet PCs. These portable devices have allowed writers to take to the streets—and, of course, to college quads, downtown coffee houses, and suburban book-stores—wherever, in short, wireless hot spots can be found.

The benefits for writers are clear, even if the connection is sometimes spotty. Writers can access documents, locate information, and communi-cate with others in a far wider range of locations. They can tap into learn-ing resources, such as the growing number of Web-based instructional writ-ing environments, whenever and wherever they need them. Most impor-tant, wireless connections allow writers to integrate these activities more fluidly into their composing processes than would be the case if they did not have immediate access to the network.

For writing instructors, in contrast, the implications of wireless connectivity are only beginning to be understood. As we've seen with other technological innovations, from word processing to hypertext to chat, early enthusiasm for wireless technologies is being tempered by the realization that it might not be all things to all writers. Preliminary (and largely informal) assessments of wireless technologies by writing instructors suggest that, as was the case with wired network access, wireless access can change the dynamic of the classroom in which it's used. Effects reported suggest that portable, networked computing devices (and, in particular, wireless laptops) allow greater flexibility in how classroom space is configured and, for that matter, where classes can meet (Allen, 1996; Bauman, 2003; Blech, 2004). Others have argued that the portability and flexibility afforded by devices such as wireless laptops support group work, small-group discussions, and project-based learning more effectively than conventional wired computers (Lowther, Ross, & Morrison, 2001; Nilson, 2003; Staley, 2004). Some educators have raised concerns about challenges to instructor authority and control (Bhave, 2002) and the relatively higher susceptibility to theft and damage of laptops, handhelds, and tablet PCs (Allen, 1996). Other scholars, however, have argued that access to portable, networked computers leads to desirable changes in the classroom (Bauman, 2003; Blech, 2004; Lower, Ross, & Morrison, 2001). Lowther, Ross, and Morrison (2001), for example, reported that their study of the use of wireless laptops in fifth- and sixth-grade classrooms indicated the use of "more student-centered strategies such as project-based learning, independent inquiry/research, teacher as coach/facilitator, and cooperative learning" (p. 7). In general, work in this area suggests that increased access to computers and computer-networks is correlated with improvements in writing skills, increases in time spent writing, increases in time spent working collaboratively, and increased flexibility in use of classroom space.

In this essay, we attempt to extend the discussion of the impact of wireless computing on writing classrooms by reporting the results of our own 17-classroom study that used wireless laptops to support instruction. Our study draws on and extends work done in an earlier study that assessed the impact of networked computers on teaching and learning in writing classrooms (Palmquist, Kiefer, Hartvigsen, & Godlew, 1998). That study, which we will refer to as the Transitions study, focused on how instructors adapted to changes in classroom context as they taught the same course in the same academic term in both traditional and computer-supported classrooms. The Transition study also explored student attitudes toward writing, student and instructor interaction, and student writing processes. Key findings included:

- Teachers adopted different roles in the two classrooms, relying more heavily on lecturing and using other front-of-the-classroom techniques in the traditional classrooms while using more student-centered activities in the computer classrooms.
- Teachers adopted different attitudes toward their students, directing students in the traditional classrooms and expecting students to take charge of their own learning in the computer classrooms.
- Students and teachers interacted more frequently in the computer classrooms than in the traditional classrooms.
- Students interacted with each other more frequently in the computer classrooms.
- Interactions among students in computer classrooms tended to focus on writing-related topics whereas interactions among students in traditional classrooms tended to be "off-task."
- Students in the computer classrooms were more confident in their writing abilities at the end of the term than at the beginning.
- In the traditional classrooms, writing was viewed as an object of study and students resisted writing during class; in the computer classrooms, writing was viewed as an essential part of class activity.

These findings suggested important differences in the nature of teaching and learning in traditional and computer-supported writing classrooms, differences that, once understood, allowed instructors to apply techniques developed for the computer classrooms to instruction in traditional classrooms. Perhaps more important, the results of the Transition study allowed instructors to gain insights into how teaching strategies, interaction, and student attitudes toward writing were shaped by the physical presence of computers and the design of the computer classrooms. As writers and instructors begin making yet another transition—this time, from the hardwired computer classroom to the potentially more flexible space of the wireless laptop classroom—we can use our understanding of previous transitions to learn about the nature of this new learning space and, by extension, to prepare more fully for teaching within it. Below, we describe the methods we used to carry out the study and discuss its results.

THE STUDY

Using the investigative framework provided by the Transitions study, we explored the instructional methods and approaches used in the wireless classrooms, the use of writing during class sessions, and interactions among students and instructors during class sessions. We gained insights into these issues through interviews with instructors and an end-of-term survey of students.

Institutional Setting

Our study was carried out at Southern Connecticut State University. Located in New Haven, "Southern" enrolls more than 12,000 students and offers 115 undergraduate and graduate degree programs. The English department has approximately 35 full-time and 35 part-time faculty members. Composition and technical writing classes are capped at 20 students, and developmental writing classes are capped at 12.

Classroom Setting

Class meetings were typically held in one of two "smart" classrooms, each equipped with a projection screen, ceiling-mounted computer projector, conventional document projector, speakers, and printer. Each room is furnished with moveable tables that fit two students to a side. Moveable chairs facilitate rearrangement of the room. At the beginning of class, students checked out one of 24 wireless Dell Inspiron laptops. Between classes, the laptops were stored in mobile, locking carts equipped with battery chargers. Between classes, the laptop carts were placed in storage closets in two of the classrooms. At the time of the study, the laptops had been in use for two years.

Instructional Technology Resources and Support

The wireless laptops used in the classrooms we studied were funded by an internal University grant awarded to the composition program (Hochman, Dean, Hood, & McEachern, 2003). Support for integrating the technology into course instruction was provided by Will Hochman, a member of the composition faculty and chair of the department's technology committee. Before the semester began, he met with instructors individually and as a group to discuss use of technology to support student composing processes, communication among students and between students and instructors, collaboration and peer review, exchange of drafts, and access to instructional and information resources. During the semester, he visited classes to provide support for instructors, distributed useful links and articles relevant to teaching with technology, and led workshops for instructors using technology in their courses. Throughout the semester, he and the participating instructors also engaged in discussions of teaching with technology over a department email list.

At Southern, all full-time instructors receive a computer of their choice (most now choose laptops) every three years. Part-time instructors are not given University computers but have the same faculty privileges of access to and storage space on the University computing network. The Office of Information Technology (OIT) at Southern oversees the use of instructional technology across the University and provides support for computer classrooms and labs as well as faculty and staff development. Although Southern does not have a comprehensive wireless access policy, OIT requires the use of standard wireless hubs and is attempting to extend the reach of its wireless network across the entire campus. OIT receives feedback on policies and procedures affecting instruction through the Educational Technology Administration committee, of which Hochman is a member.

Students at Southern can obtain information technology support through the OIT's Student Technology Resource Center. The Center's goal is to ensure that all students at Southern have access to a personal computer, network connectivity, and a basic level of software (e.g., communications, Web browser, word processing, e-mail). The Center offers tutoring in the use of popular office application software, provides access to e-mail/web kiosks and graphics/multimedia stations, and operates a computer loan program for students who cannot afford computers of their own.

The Center houses the STARS (Student Technology Assistant Representative) program. STARS students support drop-in students. They also serve as assistants for instructors who are using information technology to support their classes. Students in the STARS program are trained at the beginning of each semester to support instructors and students using the laptop classrooms.

At the request of instructors participating in the study, STARS students assisted in the classrooms, providing support to individual students and general demonstrations on how to use the laptops. Instructors reported that the most effective support from STARS students occurred within the context of student writing and research activities, where students were able to connect the operation of the laptops to their own composing processes. STARS students assisted students and instructors regularly at the beginning of the semester. Over the course of the semester, as students in the classes became more familiar with the operation of and security procedures associated with the laptops, instructors generally stopped requesting support from the STARS students.

Participants

Fifteen instructors and 278 students participated in the study. Eight of the instructors were tenured or tenure-track members of the department faculty; seven were teaching on short-term contracts. On average, the instructors who participated in the study had taught for about 15 years, but the range of experience varied widely. One instructor, teaching for the first time after completing a master's degree, had taught for two years as a graduate teaching assistant. Another instructor had taught for more than 30 years. Specializations varied as well. Four of the full-time instructors are composition specialists, two specialize in business and technical writing, one directs the department English Education program, and one is a specialist in 19th-century American literature. As a group, the instructors had either taught writing in classrooms supported by laptop or desktop computers or had expressed a strong interest in learning how to teach writing with computers.

The students were enrolled in 17 sections of first-year composition. Although no demographic data was collected, the students reflected the diverse student body at the University. Many of those attending Southern are first-generation college students from working-class backgrounds. Although most of the students attending are from the surrounding region, the school has students from more than 35 states and 50 nations. There is a 15:1 student to instructor ratio, and a male to female ratio of 1:1.8.

Instructors were recruited to the study through a general message sent to the department's email list. Students were invited to participate in the study through a brief classroom presentation, where they were told in general terms about the nature of the study and asked to consider completing a survey at the end of the term. Interested instructors and students were provided with an informed consent form. No monetary compensation was provided for participation in the study.

Information Collection

Instructor Interviews: Instructors were interviewed during the last three weeks of the semester. The instructors did not see these questions prior to the interviews. However, many of the questions addressed issues that had been discussed in presemester instructor meetings as well as in workshops conducted during the semester. The interview questions were adapted from interview questions asked in the Transitions study (Palmquist et al., 1998). Interview transcripts were analyzed to reveal trends in instructor assessments of the wireless laptop classroom as an instructional environment and in their reports of teaching strategies, lesson planning, and interaction with students. The following questions were asked:

1. What advantages, if any, do you think that the wireless laptop classroom offers as an environment to teach in over a traditional classroom?
2. What advantages, if any, do you think it has over a computer-based classroom that uses standard desktops?
3. How do you typically run a class in the laptop classroom? Do you follow a typical pattern, such as beginning with a writing activity?
4. What role does writing (drafting, revising) play in your class meetings?
5. If you use writing during class, what do you do as students write?
6. Do students ask you to help them with their writing during class writing sessions? If so, please describe some typical interactions?
7. Do students share their writing with you before or immediately following class? If so, please describe some typical interactions?
8. How do students approach writing during class? Do they see it as a natural thing to do in a writing classroom? Does their attitude toward writing in class differ from those of students you've worked with in traditional classrooms?
9. Do you feel as though you interact with students in the laptop classroom about as much, less than, or more than in a traditional classroom? If there are differences, to what do you ascribe them?
10. Do you feel you can accomplish as much in a laptop classroom as you can in a traditional classroom? How, if at all, does your planning and class management differ between the two settings?

11. Compared to your work in a traditional classroom setting, do you feel that you relate to your students similarly or differently in the laptop classroom?
12. Do you feel that you have to give up any control in a laptop classroom? If so, is that a good thing, a bad thing, or just a different thing?

Student Surveys: Students were surveyed at the end of the semester. Instructors were not informed of the results of the survey until after final course grades had been submitted and then only in aggregate across all 17 sections of the course. The survey was adapted from the student surveys used in the Transitions study (Palmquist et al., 1998). Additional questions were developed to assess student attitudes toward using wireless laptops and to assess student interaction before, during, and immediately after classes. The following items were including on the survey (see Figure 6.1):

Figure 6.1. Survey Tool

EXPECTATIONS ABOUT WRITING:	VERY LITTLE				VERY MUCH
1. In general, how much writing do you think will be required in your classes at [Southern]?	1	2	3	4	5
2. How much writing do you think you will be required to do after you graduate?	1	2	3	4	5
3. How important do you think writing will be to your career?	1	2	3	4	5
GRADES:	AGREE				AGREE
4. In this class, I expect to receive a grade of	A	B	C	D	F
5. In previous writing classes, I have usually received a grade of ...	A	B	C	D	F
ATTITUDES ABOUT WRITING:	STRONGLY DISAGREE				STRONGLY AGREE
6. Good writers are born, not made.	1	2	3	4	5
7. I avoid writing.	1	2	3	4	5

8. Some people have said, "Writing can be learned but it can't be taught." Do you believe it can be learned?	1	2	3	4	5
9. Do you believe writing can be taught?	1	2	3	4	5
10. Practice is the most important part of being a good writer.	1	2	3	4	5
11. I am able to express myself clearly in my writing.	1	2	3	4	5
12. Writing is a lot of fun.	1	2	3	4	5
13. Good teachers can help me become a better writer.	1	2	3	4	5
14. Talent is the most important part of being a good writer.	1	2	3	4	5
15. Anyone with at least average intelligence can learn to be a good writer.	1	2	3	4	5
16. I am no good at writing.	1	2	3	4	5
17. I enjoy writing.	1	2	3	4	5
18. Discussing my writing with others is an enjoyable experience.	1	2	3	4	5
19. Compared to other students, I am a good writer.	1	2	3	4	5
20. Teachers who have read my writing think I am a good writer.	1	2	3	4	5
21. Other students who have read my writing think I am a good writer.	1	2	3	4	5
22. My writing is easy to understand.	1	2	3	4	5

EXPERIENCES IN PREVIOUS WRITING CLASSES:

23. On some of my past writing assignments, I have been required to submit rough drafts of my papers.	1	2	3	4	5
24. I've taken some courses that focused primarily on spelling, grammar, and punctuation.	1	2	3	4	5
25. In previous writing classes, I've had to revise my papers.	1	2	3	4	5
26. Some of my former writing teachers were more interested in my ideas than in my spelling, punctuation, and grammar.	1	2	3	4	5

27. In some of my former writing classes, I've commented on other students' papers.	1	2	3	4	5
28. In some of my former writing classes, I spent a lot of time working in groups.	1	2	3	4	5
29. Some of my former teachers acted as though the most important part of writing was spelling, punctuation, and grammar.	1	2	3	4	5

ATTITUDES ABOUT THIS CLASS:

30. I feel comfortable asking my instructor to help me with my writing during this class.	1	2	3	4	5
31. Before, during, or immediately following this class, I have asked my instructor to look at some of my writing.	1	2	3	4	5
32. When I have worked on my writing in this class, I have asked advice from or shared my work with my classmates.	1	2	3	4	5
33. Before class sessions have formally started, I've asked my classmates to look at my writing.	1	2	3	4	5
34. I see writing (during class meetings) as a valuable part of this class.	1	2	3	4	5
35. Before, during, or immediately following this class, my instructor has talked personally with me about the work I am doing for this class.	1	2	3	4	5

ATTITUDES ABOUT WIRELESS LAPTOPS:

36. I own a wireless laptop.	No				Yes
37. I do not own a wireless laptop, but I plan to purchase one in the next year.	1	2	3	4	5
38. My instructor has done a good job of integrating wireless laptops into class work.	1	2	3	4	5
39. I wish that I had taken a writing class that does not use computers during class.	1	2	3	4	5
40. Access to a wireless laptop has helped me become a better writer.	1	2	3	4	5

Analysis focused on both individual items and subscales. Writing confidence was assessed using items 11, 16, 19, 20, 21, and 22. Writing anxiety was assessed using items 7, 12, 17, and 18. Previous writing instruction that focused on writing product was assessed using items 24, 26, and 29. Previous writing instruction that focused on writing process was assessed using items 23, 25, 27, and 28.

The measures of writing confidence and writing anxiety have proven reliable across a series of studies (Charney, Newman, & Palmquist, 1995; Mellenbacher, Miller, Covington, & Larsen 2000; Palmquist & Young, 1992). The writing anxiety questions are based on Daly and Miller's landmark studies of writing apprehension (1975a, 1975b), and the writing confidence questions were developed to parallel and extend items in the Daly Miller instrument (Palmquist & Young, 1992).

RESEARCH QUESTIONS

Six questions guided our analysis of the information we collected through the interviews and surveys:

1. How do attitudes of students who participated in this study compare to those who participated in the Transitions study?
2. How do student reports of previous writing instruction compare across the two studies?
3. How did wireless laptops affect the classroom environment?
4. How did teachers adapt their pedagogy to account for access to laptop computers?
5. What role did writing play in class sessions?
6. What sorts of interactions were reported by students and teachers?

RESULTS AND DISCUSSION

Question 1. How do attitudes of students who participated in this study compare to those who participated in the Transitions study?

In the survey, we asked students to respond to a series of statements designed to assess their writing confidence and writing anxiety. Students in the wireless laptop classrooms had a mean score of 3.49 on the writing confidence scale (a one-to-five scale; SD = 0.71, DF = 274). In contrast, at the same point in the academic term, students in the Transitions study who took courses taught in the computer classrooms had a mean score of 3.54 (SD = 0.77, DF = 78), and those who took courses taught in the traditional classrooms had a mean score of 3.33 (SD = 0.73, DF = 84). A one-way ANOVA indicated no statistically significant differences among these three groups of students.

Students in the wireless classrooms had a mean score of 2.83 on the writing anxiety scale (SD = 0.92, DF = 276). In contrast, the students in the computer classrooms in the Transitions study had a mean score of 2.73 (SD = 0.90, DF = 77), and those in the traditional classrooms had a mean score of 2.9i (SD = 0.92, DF = 84). Once again, a one-way ANOVA indicated no significant difference among the three groups of students.

Question 2. How do student reports of previous writing instruction compare across the two studies?

In the survey, we asked students to respond to items about the instruction they had received in previous writing courses. In the Transitions study, we had found that students whose previous writing instruction had not emphasized the writing process (e.g., collaboration, multiple drafts, and revision) had earned lower grades than those who were familiar with a process approach to writing instruction. This had been found, however, only for students in the computer classrooms. We suspected that this might reflect a lower level of preparation for peer review and other collaborative activities, which were emphasized more heavily in the computer classrooms than in the traditional classrooms.

An ANOVA indicates statistically significant differences among the three groups of students (F = 10.21, P < .001, DF = 2, 434) in the measure of prior instruction that emphasized product. Students in the traditional and computer sections of the Transitions study had a mean response of 3.02 (SD = 0.70) and 3.05 (SD = 0.73), respectively, and students in the wireless classrooms had a mean response of 3.35 (SD = 0.72). We found no statistically significant differences among the three groups of students in their responses to the items concerning previous instruction that emphasized process.

Question 3. How did wireless laptops affect the classroom environment?

One of the key findings in the Transitions study was that the physical presence of computers seemed to affect the behaviors and attitudes of students and instructors. In part, this was attributed to the design of the computer-supported writing classrooms, which placed a conference table at the center of the classroom and computers in an outward-facing ring around the room. Students moved back and forth between computers and discussion (in small groups or as a full class) by turning their chairs, which had rollers to facilitate movement.

In this study, we asked instructors to compare their teaching in the wireless classrooms with their teaching in traditional classrooms and computer-supported classrooms that used standard, wired desktop computers. In comparing the wireless classroom to the traditional classroom, they focused on three general issues: writing during class, Web access, and enhanced student computer literacy. Six out of the ten instructors observed that the type of writing that took place in the wireless classrooms was "more natural" and "more real world" than the writing that took place in the traditional classrooms, where students did not have access to the word processing programs they typically used to compose. One instructor commented, "Students can write in class in a more 'natural' way. Word processing has replaced pen and paper as the norm for most students." Three of the instructors noted, as well, that it was helpful to be able to view student writing during in-class writing activities. Three also commented on the advantages of being able to talk with students about their writing as they composed.

One teacher characterized in-class writing in the wireless classroom as "comfortable" and "familiar:"

> Most students compose on computers now outside of the classroom. For those of us (most of us who teach writing, I hope) who ask students to write in-class, we are able to provide students with their most comfortable and familiar writing situation. . . . In addition, the laptop classroom offers us opportunities to project the composing processes of students onto a large screen, to work on the draft as it is being composed, rather than modeling a very old-style-days-of-typewriters composing process—one in which a complete draft is completed, revised, then rewritten from start to finish. This is not the composing process students (or most of us) have anymore. In addition, students can, while working, turn to a peer for advice or reassurance that an idea/paragraph/sentence is working.

In their reflections on differences between the wireless classrooms and the wired, desktop-computer classrooms in which they had taught before, the instructors focused on the flexibility and portability of the laptops used during class, issues of visibility, and the ease of removing laptops from students' gaze. In general, their comments reflected experiences teaching in a standard, lecture-style computer classroom, in which fixed workstations are placed in rows facing the front of the classroom. Six instructors commented that the wireless laptops, which could be moved around the room—or even out of the room—allowed for a flexible arrangement and rearrangement of the classroom space. "Laptops are less intrusive, particularly in relation to the space between students," said one of the instructors. "Students are able to work very closely together, at tables, to see each other. In desktop labs, such as the foreign language lab (at least part of it), students seem more isolated, all alone in the writing process. If we want to stress the idea that writing is a collaborative, social process, the laptops help us do this."

Seven instructors also noted that, because the laptop screens could be lowered, they did away with the disrupted sight lines of the desktop computer classroom they had used in the past. "Open laptops are small enough that students can't 'hide' behind them," said another instructor. "Instructors can ask students simply to put the tops of the computers down whenever they should be engaged in a group conversation, for example, rather than working on the computer."

Question 4. How did teachers adapt their pedagogy to account for access to laptop computers?

In their interviews, teachers commented on their typical lesson plans, their use of writing in class, and issues of teacher control and student responsibility. Six of the instructors said that they did not follow a typical class plan, and four said that they always began class with some sort of writing activity. In general, however, the instructors noted that their class plans typically included some sort of writing activity—and often more than one—during a given class session. They also commented on the need to adapt lesson plans to student needs. "I fear that there's nothing typical for me," said an instructor. "Sometimes I start the day with 'get a laptop and pull up your draft,' other times we discuss goals and so on in class and then power up, other times we work in and out of researching, writing, editing, and responding. Usually my plans end up being continually revised during the class."

The need to adapt to student needs reflected a dominant theme in the instructor interviews: The shift toward a less directive style of teaching (mentioned by seven instructors) and an expectation that students should take more responsibility for their own learning (mentioned by four instructors). These comments mirror observations made by instructors in the Transitions study, where instructors seemed surprised by how differently they approached teaching in the two classroom settings. In this study, the interviews reveal that teachers tended to use less front-of-the-classroom instruction than they typically used in a traditional classroom. The interviews also suggest that they adopted a different role, using terms such as "coach" and "fellow writer" to describe how they related to their students. One instructor observed:

> I'm less likely to feel my teacher role as controlling, as up-front-professing. It could be partly that a traditional classroom reminds me of my own traditional education. The teacher professes; the student takes it in. . . . For some reason, the laptop classroom reminds me, perhaps, that we know better than all that now—that teaching and learning are about the students and not ourselves. Sometimes I'm inspired and deeply moved just to be quiet and watch them work for a minute or two. It reminds me that our students do, in fact, want to be there and do want to be working. It humbles me that they are so willing to work in front of us, as teachers. And it reminds me that I have done something right for them to risk composing in front of me—they trust in my judgment, that it will be fair, helpful, and encouraging.

Question 5. What role did writing play in class sessions?

One of the central findings of the Transitions study was that in-class writing played a significantly more important role in the computer classrooms than in the traditional classrooms. The computer classrooms, in essence, were seen as a place where writing was done. In contrast, in the traditional classrooms, writing was an object of study. It was talked about, but seldom done in class. Some of the reasons for this stemmed from the instructors' roles as teachers—in the computer classrooms, they tended to see their role as facilitators and coaches rather than as dispensers of knowledge to their students. Other reasons were rooted in student responses to in-class writing activities. On the whole, students in the traditional classes resisted writing during class sessions, while those in the computer classrooms saw it as a natural and reasonable activity.

In this study, we found parallels in instructors' attitudes toward writing, use of writing during class, and student attitudes toward in-class writing. One instructor observed, for example, that the amount of in-class writing had steadily increased over the term. "I've devoted more and more time to actually drafting and revising and editing in class, especially for freshman writers," said the instructor. "It's a good way to model the time it actually takes to write (they think 30 minutes is a long time to spend on a revision), and it is difficult for many students to have access to computers otherwise."[1]

Another instructor called attention to the wide range of writing activities that had been incorporated into class sessions:

> I often ask students to do quick writes so they can record their imme-
> diate impressions after in-class discussion, sometimes to lead to fur-
> ther discussion. Students also brainstorm their ideas on assigned essay
> topics in small groups before writing out-of-class drafts. A lot of class
> time is given to revising workshops of various kinds: working on their
> first draft by members of a small group, then writing comments for a
> peer's second draft, and finally reviewing the third draft of a member
> of another group. Students also edit and, on occasion, proofread each
> other's drafts.

Still other instructors noted that their teaching strategies had become reliant on technology. "It's at the core of what I do, and it can be easily facilitated by computers," said an instructor. "I have activities (such as using the insert comment function in Word) that specifically force students into more elaborated responses in peer workshops. Without the technology, I don't think that students would gain as much from my very structured peer reviews."

The instructors reported using a wide variety of writing activities during class sessions. Nine of the ten instructors said that they asked students to draft during class, and eight asked students to revise their work. Five of the instructors asked students to engage in writing in response to class reading or prompts, or as part of online chat or forum discussions. Seven of the instructors used in-class writing to support peer review activities.

Students' perceptions of writing during class were largely positive: 57.87% of students agreed or strongly agreed to survey item 34, *I see writing (during class meetings) as a valuable part of this class,* whereas only 14.6% disagreed or strongly disagreed. Their perception of the importance of in-class writing might have its roots in the type of interactions they had with classmates and their instructors as they wrote. One teacher observed:

> I think the writers who work in the laptop classrooms take themselves
> more seriously as writers. I think they get more of a feel for the work
> of writing as both individual and collaborative at the same time, and
> more of a sense of my role as facilitator rather than corrector, because
> of the kinds of conversations I have with them in the classroom, where
> I really strive to position them as the decision makers.

Another instructor said in an interview, "Most of my students have told me that they have never written this much before in a classroom. By the end of the semester they realize the value of extensive writing and revision."

Question 6. What sorts of interactions were reported by students and teachers?

Among the most striking finding of the Transitions study was the significantly higher levels of student-instructor and student-student interaction in the computer classrooms. In this study, which did not compare instruction by the same teachers in traditional and wireless classrooms, similar comparisons are not possible. However, the student surveys and instructor interviews provide some insights into the nature of interaction taking place in the wireless classrooms. In general, those results suggest that a dynamic similar to what was found in the computer classrooms in the Transitions study was taking place in the wireless classrooms.

The survey results indicate that students in the wireless classrooms were willing to talk with their instructors and did so regularly. Their mean response to item 30, *I feel comfortable asking my instructor to help me with my writing during this class,* was among the highest in the survey at 4.04 on a one-to-five scale (SD = 1.104, DF = 277). Their responses to survey items 31, *Before, during, or immediately following this class, I have asked my instructor to look at some of my writing,* was also strongly positive: 48.7 % of students agreed or strongly agreed and only 21.0 % disagreed or strongly disagreed. Similarly, their responses to item 34, *Before, during, or immediately following this class, my instructor has talked personally with me about the work I am doing for this class,* indicates substantial interaction with the instructor: 53.3 % of students agreed or strongly agreed and only 18.6 % disagreed or strongly disagreed.

Eight of the 10 instructors agreed that, in comparison to the amount of interaction they typically have with students in traditional classrooms, more student-instructor interactions occurred in the wireless classroom. "Students seem to associate being in the lab with hands-on activity," said one instructor. "So they abandon the attitude of mere listener and note-taker." Another instructor called attention to the important role that in-class writing activities played in encouraging interaction between students and instructors:

> I encourage [discussions with students during writing sessions] by walking around the room and peeking over shoulders or seeking out students who appear to be struggling. When I check drafts, I ask students about them and try to help them by asking them questions to

help them identify their early writing goals. Occasionally students will show me particular points in their essays and ask for help with them. Again, I try to throw questions back at them to help them make decisions about what they want their writing to do. Again, this is often easier in a laptop classroom since students can begin to make changes while we talk or immediately afterward. In regular classrooms, these discussions are often fruitful too, but students tend to forget the details by the time they get back to their dorms to revise.

In the Transitions study, students in the computer classrooms reported roughly 50% more interaction with classmates than did their peers in the traditional classrooms. In this study, student responses to the survey suggest that a similar level of interaction occurred among students in the wireless classrooms: 55.5% of students agreed or strongly agreed with item 32, *When I have worked on my writing in this class, I have asked advice from or shared my work with my classmates*, and only 16.3% disagreed or strongly disagreed. One of the instructors suggested that interactions among classmates were supported by in-class writing and peer review activities. "I've found that holding classes in the lab creates more collaborative working relations between students," said the instructor. "They're writing apprentices working on projects together. Probably this attitude derives from the opportunity for immediate hands-on writing, especially since these days students usually prefer to type rather than handwrite."

CONCLUSIONS

These findings are similar to those of the Transitions study. Notably, we saw reports of interaction among students and between students and instructors that are consistent with the findings of the Transitions study. We also saw indications that writing, as an activity, was viewed in a manner similar to that found in the computer-supported classrooms in the Transitions study. Perhaps most important, the instructors in this study showed similar attitudes toward teaching and student responsibility. Most instructors reported an emphasis on student-centered pedagogy, and several noted that they expected their students to take greater responsibility for their learning than they expected of students in traditional classrooms.

Beyond direct comparisons with the Transitions study, these findings suggest important advantages of wireless classrooms over traditional and desktop-supported writing classrooms. The instructors reported in particular on the ergonomics of wireless laptops, noting the greater flexibility they allowed in designing and carrying out instruction. The instructors referred to the ease with which laptops could be moved, covers could be lowered, and laptops could be put away. These ergonomic factors enhanced contact between students and instructors, supported collaboration among students, and reduced the extent to which, in comparison with desktop classrooms, the technology dominates the classroom space. These findings are consistent with reports published recently as part of a *Kairos* coverweb on portable technologies, in which Christopher Dean, Will Hochman, Carra Hood, and Robert McEachern (2004) called attention to the changes associated with the growing use of small, wireless laptops in writing classrooms:

> Many classroom teachers celebrate this change after years of trying to see students behind or around monitors and large boxes. Along with smaller learning stations, wireless technology is now beginning to free students and teachers who were wired into a fixed place in their computerized classes. (Why Wireless Laptops?, ¶ 1)

Few technological difficulties have been associated with using the laptops in writing classes at Southern. Early problems with running the laptops on battery power were solved by purchasing extra batteries, and occasional network problems appear to be no worse than those experienced with most university networks. Even reasonable concerns about the time associated with distributing and booting up wireless laptops at the start of each class session have been addressed by the instructors. Robert McEachern, echoing comments made by instructors during interviews for this study, suggested that time spent distributing laptops can be viewed as time spent bringing students and instructors into contact:

> Ten or 20 seconds seems pretty insignificant. But I'll ask you—how many classes have you taught where you greet each student by name before class starts? I don't have evidence that this is contributing to better writing, or happier students. But it feels good to me. I'm assuming it feels good to students. (Dean et al., 2004, My New Reality, ¶ 5)

Anecdotal evidence suggests that a significant number of the faculty members at Southern view the changes associated with teaching in wireless laptop classrooms as beneficial to their teaching. In the two years since the laptops were made available, only one instructor who has taught with them has decided not to use them again. Moreover, every slot in both of the classrooms is eagerly sought by instructors, and a third classroom is being developed.

Even so, the number of instructors who have asked to use the laptops is still less than half of the total department faculty. And concerns have been expressed by faculty who view with suspicion the use of technology to support writing instruction. As Dean et al. (2004) noted in their article in *Kairos*, this suspicion seems to be associated with the desire to hold onto print culture even as we move into a hybrid stage of print/digital writing. Reflecting on his initial resistance to teaching with computers, Christopher Dean observed:

> I now know that my [initial] resistance is in some ways justified, and the principle way in which I justify it is based on the work of Cynthia Selfe, particularly her piece "Redefining Literacy: The Multilayered Grammars of Computers." This piece claims that multilayered literacy is a literacy in which people "function literately within computer-supported communication environments" by layering "conventions of the page and conventions of the screen" (7-8). As Selfe argues, writers inhabit this space—a space in which the hum of electronic discourse is accompanied by the scratching of the pen (7-8). Selfe seems to think that this is a transitional stage, but I would argue that teachers should continue to think about student reading, writing, and academic achievement in relation to the intersection of computer and page for many years to come. (Dean et al., 2004, The Many Colored Coat of the Emperor, ¶ 11)

No matter how long the transition lasts—or whether or not it ultimately proves to be a transition in any sense of the word—this study suggests that the primary challenge faced by instructors who use wireless laptops is not technological. Instead, as the instructors we worked with in this study and in the original Transitions study indicate, the most important challenge is pedagogical. One element of this challenge is how our students conceptualize a learning space. In the not too distant past, most writing students' conception of a learning space was a classroom with rows of desks. In the late 1980s and 1990s, we witnessed a widening of that conception as students began to learn to write in rooms containing networked computers, moveable tables, and chairs. In the near future, our students are likely to begin thinking of a learning space as anywhere they might connect to the network. This evolving sense of where—and how—one learns is mirrored in the growing ease with which students can move and access information. As students come to rely more and more on flash drives, network storage, and

Web-based learning environments, we'll see a shift from a brick-and-mortar notion of classrooms to one in which wireless networks are as much a part of the classroom space as overhead projectors and white boards and where students view easy access to course resources outside of the classroom space as something as natural as buying textbooks and writing essays.

Another element of this challenge is how instructors will conceptualize the place of wireless networks in their courses. The results of our study suggest that the use of wireless laptops shape the conduct of class and the nature of classroom activities. These ergonomic effects are important, but they reflect early efforts by instructors to integrate wireless networks into their courses. As instructors become more familiar with the characteristics of these networks—and as the networks and the devices that connect to them become more powerful—we should expect additional affects on how writing is learned and taught. What those effects will be is unclear. They will no doubt reflect changes in higher education as a whole. They will also reflect our understanding of what it means to be a writer, how writers share their work, and the shapes of the work they share. What seems clear, however, is that even classrooms that outwardly resemble the 19th-century, rows-of-desks configuration will be affected by instructors' understanding of how to use wireless networks to extend the boundaries of the classroom.

As writing instructors and program administrators, we need to consider not only what is made possible by the technology we use in our classrooms—from chalk to overhead projectors to wireless laptops—but also what is obscured and what is made difficult. As Dean suggests, our hybrid culture is here to stay, and it is clear that our pedagogies have not always kept pace with the technological changes that are shaping our classrooms and our teaching and learning lives.

As writing technologies continue to become more transparent, we will move from questioning what it is to investigating how we can best live and learn with it. In one sense, it's likely that this latest shift in the context of our learning spaces will be easier to address than some earlier shifts. The physical intrusion of computers into classrooms is something we've grown used to. If anything, this shift will minimize that intrusion, as students and instructors come to rely on smaller, less intrusive digital tools—one day perhaps on devices that use digital paper and flexible screens as input and output devices. Rather than worrying how we can equip classrooms with the latest technology, we'll begin expecting students to bring that technology with them. In a sense, for writing program administrators, this might be the easiest shift of all. For curriculum designers and teachers, however, the challenge will be investigating what has become possible and whether—and if so, how—it should be integrated into instruction.

This study and the earlier Transitions study emphasize the ongoing need for pedagogical direction and reflection. If we can adroitly learn to connect our educational goals to useful changes in technology, we might also be able to use this process of hybrid adaptation as an object lesson for our learning processes in general.

ENDNOTES

1. Even though students could obtain access to laptops through the Student Technology Resource Center's computer loan program, which was housed in the library, a number of students chose not to take advantage of this resource. One instructor noted, "The difficulties of encouraging students to use library laptops are sometimes similar to getting them to use library books."

REFERENCES

Allen, Nancy. (1996). Designing an electronic writing classroom. *IEEE Transactions on Professional Communication, 39*, 232-238.

Bauman, Marcy. (2003, December 5). Re: [techrhet] need help with hoped for wireless lab. Message posted to TechRhet electronic mailing list. Retrieved March 18, 2005, from http://www.interversity.org/lists/techrhet/archives.php

Bhave, Mahesh P. (2002). Classrooms with Wi-Fi. *T H E Journal, 30*(4). Retrieved January 23, 2005, from Academic Search Premier database.

Blech, Bradley. (2004, March 5). Re: [techrhet] Help! Message posted to TechRhet electronic mailing list. Retrieved March 18, 2005, from http://www.interversity.org/lists/techrhet/archives.php

Charney, Davida, Newman, John H., & Palmquist, Mike. (1995). "I'm just no good at writing": Epistemological style and attitudes toward writing. *Written Communication, 12*, 298-329.

Daly, John A., & Miller, Michael D. (1975a). The empirical development of an instrument to measure writing apprehension. *Research in the Teaching of English, 9*, 242-249.

Daly, John A., & Miller, Michael D. (1975b). Further studies in writing apprehension: SAT scores, success expectations, willingness to take advanced courses and sex differences. *Research in the Teaching of English, 9*, 250-256.

Dean, Christopher, Hochman, Will, Hood, Carra, & McEachern, Robert. (2004). Fashioning the emperor's new clothes: Emerging pedagogy and practices of turning wireless laptops into classroom literacy stations. *Kairos, 9*(1). Retrieved

March 20, 2005, from http://english.ttu.edu/kairos/9.1/binder2.html?cover-web/hochman_et_al/intro.html

Hochman, Will, Dean, Christopher, Hood, Carra, & McEachern, Robert. (2003). Proposal for computerized writing instruction at SCSU. Southern Connecticut State University. Retrieved March 20, 2005, from http://www.southernct.edu/~hochman/Laptoplabproposal

Lowther, Deborah L., Ross, Steven M., & Morrison, Gary R. (2001). Evaluation of a laptop program: Successes and recommendations. In *Proceedings of National Educational Computing Conference*, Chicago, Illinois, 1-8.

Mehlenbacher, Brad, Miller, Carolyn R., Covington, David, & Larsen, Jamie S. (2000). Active and interactive learning online: A comparison of Web-based and conventional writing classes. *IEEE Transactions on Professional Communication, 43*(2), 166-184.

Nilson, Linda B. (2003). *Teaching at its best: A research-based resource for college instructors* (2nd ed.). Bolton, MA: Anker Publishing.

Palmquist, Mike, Kiefer, Kate, Hartvigsen, James, & Godlew, Barbara. (1998). *Transitions: Teaching writing in computer-supported and traditional classrooms.* Greenwich, CT: Ablex.

Palmquist, Mike, & Young, Richard E. (1992). The notion of giftedness and student expectations about writing. *Written Communication, 9*, 137-168.

Staley, David J. (2004). Adopting digital technologies in the classroom: 10 assessment questions. *Educause Quarterly, 3*, 20-26.

7

CHANGING THE GROUND OF
GRADUATE EDUCATION

WIRELESS LAPTOPS BRING STABILITY, NOT MOBILITY,
TO GRADUATE TEACHING ASSISTANTS

Kevin Brooks

All situations comprise an area of attention (figure) and a very much
larger area of inattention (ground). The two continually coerce and play
with each other across a common outline or boundary or interval that
serves to define both simultaneously. The shape of one conforms
exactly to the shape of the other. Figures rise out of, and recede back
into, ground, which is configurational and comprises all other available
figures at once.

Marshall McLuhan and Eric McLuhan,
The Laws of Media (1988)

Wireless networking and wireless devices are currently prominent figures
in relation to the wired educational ground most of us inhabit. As Marshall
McLuhan and Eric McLuhan (1988) define figure in the epigraph, wireless
technologies have become a significant area of attention, but wireless networking will recede into the ground, and in the near future, we will not likely make much, if any, distinction between our wireless and wired networks.
McLuhan and McLuhan use figure/ground as a perceptual tool primarily to
understand the "sensory bias imposed upon us by our extensions" (p. 10),
but figure/ground is also a useful perceptual tool for examining the material arrangement and prioritization of resources in media ecologies, includ-

ing the specific computing environments of our universities, departments, and graduate programs.

Existing scholarship and speculation about the role of wireless computing in higher education, like the clusters of essays from the University of North Carolina, Chapel Hill (Anderson, Brown, Taylor, & Wymer, 2002) and Southern Connecticut State University (Dean, Hochman, Hood, & McEachern, 2004), has focused on potential changes to the traditional classroom because the classroom is the most obvious space of education, and it is the space that has been the most difficult to wire without complete transformation. Scholarship on Graduate Teaching Assistants (GTAs) and their use of computer technology has focused on computer classroom pedagogy (Duffelmeyer, 2003) and classroom strategies (Meeks, 2004). In fact, many chapters in this collection describe changing classrooms, changing students, and the changing notions of composition, but the impact of a wireless computing environment on teachers receives less attention. Discussions of teachers and wireless technologies tend to focus on teachers in the classroom rather than in their offices or other spaces where they must manage data across computers and locations.

In this chapter, I focus on the behind-the-scenes transformation of the computing environment in the M.A. program in English at North Dakota State University (NDSU). I examine the impact of providing NDSU GTAs with wireless networking and wireless-ready laptops on a campus that otherwise did not support wireless computing at the time of our implementation. The English department received sufficient funding to purchase 20 Dell Latitude 505 wireless laptops and appropriate software, three institutionally installed access points (as opposed to the small, portable routers many of us use in our homes) as part of a secure, wireless infrastructure that is increasingly being deployed on our campus, and other supplies. These technology acquisitions created a 1:1 GTA-to-computer ratio, replacing the 5:1 ratio that had been the norm for the previous seven years. Not all institutions or departments will have the funds to make this transformation immediately, but English departments are often in strong ethical and pedagogical positions to argue for funding that will make this kind of change possible. English departments could wait for wireless computing to become the ground—to be become the common environment in which we work and teach—but by doing so, we would miss out on an opportunity to study, understand, and shape the future uses of these technologies. Waiting for wireless networking to become the ground and wireless laptops to become the norm will likely result in the costs getting passed on to GTAs— the last constituency who should have to pay for these technologies.

This chapter will elaborate on the establishment and execution of a wireless laptop initiative for English GTAs in the fall of 2004, as well as describe additional developments in 2005. The initiative was informed by the ethical imperative of conducting institutional critique (Porter, Sullivan, Blythe, Grabill, & Miles, 2000) and the importance of paying attention to the intersections of technology and literacy (Selfe, 1999). I will draw on McLuhan and McLuhan's notion of figure/ground throughout the essay, but the GTAs' stories constitute the heart of this chapter. In October 2004, all 19 GTAs voluntarily responded to a four-question, open-ended survey on their wireless laptop use, and in January 2005, 16 of the 19 GTAs responded to a follow-up 30-question survey on their wireless laptop use.[1] Following the suggestion of Porter et al. (2000) to conduct institutional critique through narratives that blend theory and research (p. 631), I will construct stories of GTAs' experiences from this information rather than report research findings. In my role as Writing Program Administrator (WPA), I wanted to better understand the ways wireless laptops influenced the lives of our GTAs. The GTAs' experiences suggest that wireless laptops have created a stable computing environment for their teaching and scholarship, with mobility being less important than stability. In 2005, the changes, improvements, and extensions the department and Writing Program have made to the laptop initiative have focused on ensuring the stability rather than increasing the mobility of GTAs' computing environment. Wireless technologies and the mobile life (m-life) will undoubtedly remain a prominent figure in the media and on our campuses for the next few years, but neither the technology nor the lifestyle is likely to live up to its press billings without stable infrastructures, stable devices, and meaningful uses for mobility.

INSTITUTIONAL CRITIQUE AND REDIRECTED INVESTMENT

In the fall of 2003, NDSU's English department employed 19 GTAs who shared offices with four, six, and seven people per office and with two other GTAs sharing an office with adjunct instructors. Each office was equipped with one or two desktop computers, but those computers were at best three years old, some as many as seven years old—slow and unreliable. In their essay "Institutional Critique: A Rhetorical Methodology for Change," James Porter, Patricia Sullivan, Stuart Blythe, Jeffrey Grabill, and Libby Miles (2000) challenge the field of rhetoric and writing to conduct research that will do more than "recommend or hope for institutional change" (p. 628). They suggest that those interested in a rhetorical methodology of institutional change look beyond the classroom to recognize the ways in which classroom interactions and spaces are shaped by larger institutional forces (p. 632). As I prepared to take on the WPA position and the responsibility of training new teaching assistants within my department, I realized that one of the most important things I could do to support the professional development of our GTAs would be to improve their access to computers within their work spaces.

Mark Aune, Chair of our Graduate Studies Committee, and I wrote an internal grant to our university's Technology Fee Advisory Committee (TFAC), a committee that reviews proposals to distribute a portion of NDSU's Student Technology Fee for student-centered technology projects. I had served on the committee for two years and was still on the committee in 2003-2004, so I knew that members did not like to fund desktops or laptops as part of proposals. The NDSU computing environment, however, was starting to undergo changes. Our campus had just installed its first Information Technology Services (ITS)–sponsored wireless hub in the summer of 2003. TFAC had reviewed and rejected a very large proposal for wireless networking, but the rejection was based primarily on the concern that NDSU did not have a campus-wide plan for establishing a wireless network. An Ad Hoc Wireless Task Force was formed, and we met four times in the spring of 2004 to assess and recommend a direction for developing a secure wireless network on campus, one that would support teaching and learning. Wireless computing was very much the visible technology figure on our campus in 2003-2004, as it has been on many campuses the past few years, and Aune and I hoped to make the most of this "area of attention" by seeking funding for a laptop initiative.

Our initial request in the fall of 2003 for approximately $30,000 and complete funding for laptops, software, and networking was rejected both because we had not fully worked out the wireless implementation with the ITS department and because the TFAC was reluctant to purchase computers for a department, articulating the view that departments should be responsible for meeting their own basic needs. We revised the proposal for the spring review, requested only half as much money with the goal of getting a 2:1 computer-to-GTA ratio, and this time, we received written support from ITS. Our proposal was approved, partially on the basis of our revisions and partially on the basis of the strong support by the Graduate Dean who saw our initiative as a potential model for other graduate programs facing similar space and technology access problems. After we received these funds, the Provost/Vice President for Academic Affairs contributed an additional $12,800 and we were able to achieve a 1:1 ratio. The College Dean contributed an additional $3,500 to pay for the wireless infrastructure that has the potential to benefit other departments in our building and College. We ended up with a total budget of $30,000, of which we spent $23,000 on 20 Dell Latitude 505 wireless and Ethernet-ready laptops. We also purchased Microsoft OFFICE XP PROFESSIONAL licenses for each laptop, carrying cases with shoulder straps (not backpack style), and three networked wireless access points in order to maintain fast and reliable access within GTAs' offices. We had approximately $1,500 left at the start of the fall semester to cover unanticipated costs and increased office costs—particularly paper and printer maintenance.

Because our initiative works at the intersection of technology and literacy, Cynthia L. Selfe's call for the importance of paying attention to the ways in which technology and literacy are intertwined in the 21st century resonated in our situation. Selfe (1999) notes that the "rhetoric of technological boon"—or what I call "the rhetoric of affordability and ubiquity"—obscures the real economic situation of graduate and undergraduate students (p. 26). She also explains that educators need to pay attention to the whole system of investment that our institutions, communities, and culture are making in computing technologies (p. 38). By not overestimating our GTAs' limited financial resources and coming to an understanding of our institution's system of investment through my involvement with the TFAC, our department's proposal was able to shift that system of investment in our favor.

Our laptop proposal met with initial resistance because some members of the TFAC felt that our graduate students should simply purchase their own laptops, the argument being that laptops are now affordable and ubiquitous. Some TFAC members thought that the English Department should purchase these laptops out of its operating budget, not realizing that kind of bare-bones operating budget our department, like most humanities departments in the country, operate with. This same committee, however, annually distributes between $700,000 and $800,000 to fund a variety of projects, some that impact a small number of students and others that impact most of the campus. Although we certainly did not feel entitled to TFAC funding, we had to counter the rhetoric of affordability and ubiquity. We argued that our GTAs teach four sections of first-year English per year for a very small stipend, and that this investment of student technology fee money was going to benefit both graduate and undergraduate students. We had to convince TFAC that the old patterns of investment that the committee typically made need not constrain new investments—laptops for individuals—and that a new kind of investment in graduate students would also be an investment in undergraduate students' learning experience. And finally, we had to convince the proposal reviewers that literacy and technology are indeed closely intertwined, and that literacy instruction provided by GTAs with limited access to the now-basic tools of teaching and communicating did not make good pedagogical sense.

In writing our proposal, we were aware that we were trying to take advantage of the dominant figure of campus computing in 2003-2004—wireless networking. But we were also trying to pay as much attention to the ground as possible, which meant things like working closely with the ITS department for support, including funds in our budget for increased paper use, and identifying areas of training and pedagogy that we would need to address with our GTAs. We were expecting that the real payoff of the wireless laptop initiative would actually come from the 1:1 GTA to computer ratio, more so than the wireless connectivity. We were convinced that providing our GTAs the technology to prepare effectively for class and communicate with their students would have more significant impacts than making our GTAs more mobile. Their experiences in the fall of 2004 supported many of our assumptions, but the GTAs also showed us ways in which the wireless laptop initiative created benefits and physical constraints we had not anticipated.

THE GTA EXPERIENCE: MOBILITY OVERRATED, STABILITY UNDERRATED

In this section I draw on the two surveys of GTA laptop experiences. Due to space constraints and my own desire to craft a local story of institutional critique, I present the results as impressions rather than quantified results; this data helps me explain some of our new understandings about sustainable computing. I acknowledge, however, that our focus on stability and mobility by no means tells the whole story of our wireless laptop initiative. Instead, it is a story of figure/ground, a story of paying attention to the environment around us rather than the technologies that catch our eyes.

Mobility Overrated

The expected outcome of wireless connectivity is mobility, according to mass media representations such as the Best Buy™ Pinocchio ads in which the wires are finally cut, leading to uninhibited computing. Of the 19 students who responded to the midterm survey questions, however, only six explicitly identified mobility or portability as a major benefit of having a wireless laptop. One student wrote about his mobility on campus and at home.

> I can pick up a wireless connection in my TA Strategies class . . . which allows us to assign a "surfer" during a class period to search [the Web] for information we are discussing. I can pick up wireless connections in most of the locations I frequent in Minard [Hall], making the Internet a consistent resource. Finally, there is the simple luxury of being able to sit down in an easy chair, computer in my lap and book in my hand, or outside on the front step on a nice evening. Freedom of movement is very much underrated!

Other students noted similar kinds of mobility within their homes as a benefit of having a laptop, but only one student wrote about mobility and travel during the school year.

These stories of mobility, however, are more of an exception than a rule. The end-of-semester survey found that 14 of 16 students identified their office as the primary location where they used wireless connectivity, and seven of the 16 did not use wireless connectivity off campus. When asked if being given a laptop "sent the implicit message that being a graduate student and teaching assistants would be an anytime, anywhere occupation," only six agreed or strongly agreed, and ten were neutral or strongly disagreed. Students generally did not take their laptops to the classes they taught because few if any rooms on campus had wireless access, and those GTAs who wanted to use multimedia resources either relied on their already outfitted classrooms or the institutionally available multimedia carts. The GTAs generally did not perceive mobility to be the implied message of the wireless laptops nor did they find the portability of the laptops with wireless connectivity to be their strongest feature.

No one complained about being given a laptop, but two GTAs at midsemester said that they were disappointed in the lack of mobility because there were so few access points on campus in the fall of 2004, and six students specifically reported some back, neck, and shoulder pain that they attributed to carrying the computer plus books. One student wrote:

> The main drawback for me is that my laptop isn't as mobile as I was hoping it would be. Granted, it is smaller than many of the laptops on the market, but with my slight frame, my shoulders can't support the weight of a laptop/wires/case and a briefcase/papers/books comfort-

ably. After a week of toting it around, I developed noticeable back and shoulder/neck pain. Therefore, I am only taking my laptop home when absolutely necessary.

When asked more explicitly about the physical strains of (a) carrying the computer, (b) using the keyboard, or (c) eye strain, 11 of 16 identified carrying the computer as a problem, six had difficulty using the keyboard, and nine said they suffered from eye strain—respondents could rank all that applied. The laptops functioned primarily as desktops for some students, particularly those who had a home computer with additional software and features, but 12 of the 16 respondents identified the laptop as their primary computer. As I will suggest, our wireless laptop initiative has in fact highlighted the ways in which a robust and stable computing environment can enhance GTAs' pedagogical and professional development. It is important for administrators to both consider GTA technology needs and understand that laptops may support some mobile computing, just not the magical freedom invoked by the Best Buy™ Pinocchio ad and other popular stories.

Stability Underrated

Mobility may be the image that is selling wireless, but stability was the more common benefit identified by the GTAs. Nine of the 19 midsemester respondents said that having their own laptop allowed them to use their assigned office space, cramped as it was, more effectively than spaces such as home or public computer clusters. One GTA illustrated this point clearly:

> The benefit of having the laptop is being able to use a computer during my office hours. Before, I rarely/never got to use the computer when sharing an office. It's so nice to be able to actually answer student e-mails during office hours instead of having to go to the IACC [NDSU's major computing facility at] or library.

When asked in the survey if "Having my own laptop allows me to use office time efficiently," 15 of 16 agreed or strongly agreed. The GTAs also acknowledged that having a laptop allowed them to use other locations efficiently, but our campus and their homes were used much more heavily than commercial spaces such as malls and coffee shops or others' homes. When preparing the laptop initiative proposal, Aune and I envisioned a stable computing environment where all GTAs would have easy access to reliable computers and connections. The GTAs helped extend our definition of "stable computing environment." Ten of the 19 who responded to the open-ended questions identified some version of "having everything in one place" as a central benefit. Even though having their files on a single laptop supports student mobility, these same respondents noted that they had already been mobile. It was not a new mobility they had gained but rather a new file management program. Instead of having data dispersed across home and office computers, GTAs were better able to manage their files. The laptops in these cases do not seem to have increased individuals' mobility, but the portable computer brought some coherence and stability to already mobile lives. On this point, a second-year GTA explains:

> The laptop has meant a significant change in my work. I have a desktop, and that is where I used to work half of the time the first year as a TA. I also worked with the office desktop, which meant that I had my work spread in two different places. Although I tried to have everything in floppies, it was hard to keep them updated. The laptop has solved that problem beautifully: now I have everything in the same computer, and I don't have to worry if I don't remember in which computer I have the file I need.

Because these issues of stability and mobility emerged so clearly in the midsemester feedback, I asked GTAs in the follow-up survey if having the laptop created a stable computing environment for them, enhanced an existing stable environment, or destabilized their computing environment, and not surprisingly, they all said the laptops created or enhanced a stable computing environment.

When asked in the survey to describe their computer use, 13 identified class preparation as their primary use, and three identified their own graduate studies as their primary use, with preparation as their secondary use. Only one student identified entertainment as a more frequent use than one of the other two options—the fact that their supervisor posed the question undoubtedly influenced their answers. When asked how many hours a week they used the NDSU provided computer, their home computer, or other computers, only four used other computers more frequently than their laptop, and 11 of the remaining 12 used their computers ranging from 10 to 40 hours a week—most in the 10 to 20 hour range. When asked to break down their actual use of the computer while in their offices, class preparation was still the most frequent label used, with answering email, using our course management system, and doing their own work as the three next most popular uses. All of these uses do not necessarily confirm a stable computing environment, but they suggest that our GTAs have, by and large, used their laptops to support effective teaching and completion of graduate work.

I have been arguing that stability is a good thing, but stability can also mean staleness, lack of interaction, and/or a decline of community. The midsemester feedback I received provided two opposing perspectives on the kind of community that the laptops seemed to encourage. One GTA was excited about the sense of online community and resource-sharing that seemed to be emerging because of the easy access to computers and the Web, whereas another GTA worried that the camaraderie was slipping: "I feel glued to the damn thing! Sometimes I think I'm developing an unnatural and strange relationship with my laptop . . . I think laptop use may have decreased the sociability of the graduate students." When asked in the survey if the laptops had led to some isolation or "cubicalization" of their graduate school experience, three agreed but the rest disagreed. Chris Anson (1999) articulated concerns about cubicalization (p. 269), as well as a concern about turning departments into "ghost haunts" as writing instruction becomes increasingly a computer-mediated experience (p. 270), but only two GTAs reported spending fewer than five hours a week in their office space, six reported spending 5-to-9 hours, four reported 10-14, three reported 15-19, and one reported over 20 hours per week. One semester into the wireless laptop initiative, the graduate student community was still strong and GTAs' use of office space seemed to increase, although we have no baseline for comparison.

Joshua Hernandez, a GTA who spent two years in the NDSU graduate program, delivered a paper provocatively entitled "The Gift of Fire" at the 2005 Great Plains Alliance for Computers and Writing Conference. His paper tells two contrasting stories of working in the shared office space—first wired with two desktops, then wireless with six laptops. The first story, drawn from Fall 2003, describes a day when Hernandez needed to use one of the two computers in his office just 15 minutes before class, only to find both occupied, and two other computers in the general office space also occupied. When he finally elbowed his way onto one of the computers in his office, it crashed. His analysis of the event emphasized a chaotic and competitive work environment: "Computer access was a matter of survival of the fittest. If you were on a machine before anyone else, it was yours. Going to the office to get work done was always a crapshoot, unless you went late at night when you were guaranteed computer access." The second story describes an alternative universe(ity)—life with a wireless laptop—in which Hernandez waltzes into the office at 7:45 a.m., again just 15 minutes before his class. He checks e-mail from two students, reads a departmental announcement, posts an assignment to the course management system, prints a handout, and then walks out ten minutes later. His analysis describes a different office environment than the one from his first semester.

> In-office productivity has increased and competition for computer time has been eliminated. The most noticeable effect of wireless technology and the laptop initiative has been the creation of a stable work environment for TAs. Mobility is a minor and often limited benefit. Individual laptops and wireless connectivity are in constant use during office hours. People are typing up things for their classes, or writing texts for their graduate courses. There are complaints that students e-mail us too often about trivial matters. But we respond to them anyway. Sometimes, we don't even bother taking our laptops home with us. Occasionally, it's because we prefer to work on a desktop at home; sometimes it's because we won't be needing it—we finished our computer dependent tasks for the day while in the office; sometimes we just want to get away from the damned things.

The wireless laptops have filled a basic need in our graduate program—a huge need that almost all graduate programs in English should similarly be looking to meet. Even on a campus without a wireless canopy, our small efforts have made the cramped office quarters more productive and professional. The GTAs provide their own critical analysis of the technology and the initiative—getting away from the laptops to keep a sense of balance in their lives, consciously working to maintain a rich human community in the department, and taking on reasonable and manageable changes in their pedagogy appropriate for GTAs early in their careers. The GTAs also provided their own form of institutional critique through their responses to survey questions, giving us a sense of how the department needs to further stabilize the ground of graduate education computing while we are still thinking about and focusing on the figure of wireless laptops.

ENRICHING THE ENVIRONMENT, SEEKING SUSTAINABILITY

Our initiative has room to improve, stabilize, and enrich the computing environment by improving our GTAs' pedagogical development, increasing opportunities for effectively working with wireless instructional technologies, and extending and sustaining the hardware central to the program.

This section describes developments and adjustments to the laptop initiative in the fall of 2005, drawing on a mixture of GTA experiences, events that have transpired, and actual changes and purchases we have made within the department.

Barb Blakely Duffelmeyer (2003) has written about how computer-specific pedagogy can add "an additional and almost arbitrary layer of stress and uncertainty to an already stressful and uncertain teaching experience" (p. 296), and that new GTAs will develop both their skills and "critical technological literacy" within their own communities of practice (p. 306). The end-of-semester survey results found that 15 of 16 GTAs thought the basic training on the laptops and the training related to their graduate work was adequate, but only 9 of 16 felt that their training relevant to teaching was adequate. I had organized one full workshop day prior to the semester, a day that included some basic course management software instruction as well as sessions on pedagogical uses of blogs, Photoshop, and information management. In my graduate course, I relied on the "Teaching with Technology" section in Duane Roen, Veronica Pantoja, Lauren Yena, Susan K. Miller, and Eric Waggoner's (2003) *Strategies for Teaching First-year Composition* to guide our discussion, but most students found it of limited use. We discussed and practiced peer review online, and I suggested, but did not require, the GTAs to teach a "PowerPoint Music Video" assignment that would have made significant use of their laptops and technological literacy. Duffelmeyer's observation that teaching with technology might seem like "an additional and almost arbitrary layer of stress and uncertainty" probably applied to me, more than the GTAs, as I taught our department's composition pedagogy class for the first time and was nervous about leading the GTAs into technological tangles that might discourage them from teaching with technology in the future. Despite claiming to have some of the highest technology related anxieties, two GTAs used the PowerPoint Music Video assignment. They found that our institutional resources and their own students' willingness to compose new media resulted in a successful teaching/learning experience. Such an assignment does not foreground wireless technologies, but the collecting and playing of 44 music videos would have been almost inconceivable if the GTAs did not have their own computers. Now that the GTAs have a more stable computing environment than in previous years, and now that the campus as whole is moving towards a wireless canopy, it will be my responsibility as WPA to strengthen our first-year writing program's understanding of pedagogies appropriate to and effective in a stable wireless environment.

Our GTAs are constrained in their pedagogical use of wireless technology because there are currently so few classrooms with reliable wireless connections, and even if GTAs teach in a classroom where a wireless signal is available, NDSU students still do not typically bring laptops to class. Only fully instrumented classrooms—classrooms with a desktop computer, VCR, and document reader—have projectors, making the laptop redundant if a GTA were to teach in such a classroom. To try and improve access to a truly wireless classroom, in which instructors and students have wireless laptops and a reliable signal, the ability to project from laptops, the flexibility to work individually, in small groups, or in large groups, our Department Head, Dr. Dale Sullivan, wrote a $30,000 request to our TFAC committee for a laptop cart—20 Dell Latitude 610s and appropriate software. The proposal was fully funded on first review, based largely on the fact that we have learned to work closely with our ITS department, we again secured the support of the Provost for classroom renovations, and we again featured wireless as the prominent figure in our proposal, even though we understand the whole concept—or ground—of the flexible classroom to be the most important feature of the proposal. Our wireless and flexible classroom was scheduled to be operational for Fall 2006, we plan to have GTAs regularly teach in the room, and GTAs will eventually teach hybrid classes—once a week face-to-face, once-a-week online—to extend their pedagogical training, to give more students and teachers access to the wireless classroom, and to relieve some overcrowding of classrooms on campus. Neither the wireless flexible classroom, nor hybrid classes, would be as viable or valuable if our GTAs did not have reliable, regular access to stable computing.

Although Richard Selfe (2005) recommends having a refresh plan in mind for any computer initiative (p. 107), we admittedly have embarked on the laptop initiative for our GTAs with a shaky long-term plan. We think we can refresh the laptops by staggering purchases with our annual equipment requests, and we have also re-allocated some operating funds to take advantage of sales on 12" iBook G4s, a purchase that was also informed by GTA complaints about the weight and bulk of the Dell 505s. These purchases have introduced a second type of computer into the initiative, creating as-yet-unanswered questions about viability, stability, and institutional support for our laptop initiative. The implications of our new spending priorities surfaced at a meeting when some faculty members acknowledged that all the new equipment was nice, but they wondered what could be done to improve our limited library holdings.

In an attempt to keep costs down with our initial proposal, we also relearned the age-old lesson that cutting corners has its own set of costs. In choosing the features for the Dell 505, we saved approximately $200 per machine by choosing a CD drive, rather than CD or DVD burner, although we will choose laptops with DVD players/burners when we start to replace these laptops. The GTAs in 2004-2005 had to rely on floppies to save and move files or purchase their own portable storage device. When the department made its technology purchases for 2005-2006, however, we purchased a single Dell DVD player/burner that GTAs will be able to check out and use for class or for backing-up of material (the Dell laptops can use floppy drives, CD drives, and DVD drives in the same slot), and we purchased 20 USB drives (512 MB) to provide GTAs with better back-up and file-sharing capabilities. The department also purchased a Dell portable projector that enables GTAs to take laptops to classrooms that are not outfitted or do not have easy access to multimedia carts. All of these expenditures to stabilize our initiative—the replaceable DVD drive, the USB drives, the portable projector—were within our department's budget possibilities. Only the initial big expenditure—the 20 laptops plus software and wireless infrastructure—were beyond our capabilities.

Unexpected costs, of course, could complicate and destabilize our computing environment. Since the fall of 2004, we have had to replace two hard drives on the laptops, but both of those replacements were under warranty. In the fall of 2005, we had our first major replacement: A cup of coffee was spilled on a laptop, and the repair costs would have exceeded replacement costs. All GTAs have signed a computer use agreement, which includes vague language about being responsible for the laptop in case of loss or significant repair, but this incident forced us to be more specific about how much responsibility and liability we as a department could take on, and how much responsibility and liability the GTAs would need to assume. We decided that from the point of purchase, the department would take responsibility for approximately half the cost of replacement, and with each semester the laptop is in use, the GTA's responsibility is reduced. A laptop lost or significantly damaged in its first semester of use, for example, would require a $600 repayment on the part of a GTA, but a laptop lost or damaged in its fourth semester of use would require a $200 repayment on the part of the GTA. After three full years of use, or six semesters, the laptop will no longer carry any replacement value.

WIRELESS LAPTOPS AND THEIR MESSAGES

As long as wireless networking remains a prominent figure in higher education computing, the bulk of speculation and experimentation with its uses will likely be in classroom spaces or related to extending the classroom. But wireless networking can be used as a figure to influence and strengthen the ground of the basic computing environment for GTAs in English. Within English departments and programs, we should prioritize the development of stable computing environments for all teaching staff (Selfe, 2005, p. 25). Even as laptop costs come down, handheld PC and portable storage devices increase in capabilities, and wireless networking recedes into the ground of higher education computing, it is imperative that institutions continue to do their best to provide GTAs with a stable ground of computing, whether wired, wireless, or a mix of both. Pushing the costs of computing onto graduate students will only further an already economically unbalanced situation, and introducing a wide range of computing devices through personal rather than programmatic choice may further destabilize the computing environment. When diverse devices and software are used, access to service and support may be limited, and when individuals choose their own devices, the computing power of personal devices may not be adequate for teaching and learning in 21st century electronic environments. A campus's computing network is potentially at risk if GTAs are working with equipment that does not regularly update virus protection.

As the NDSU campus increasingly goes wireless, our writing program and the GTAs who teach in it will be able to employ some of the pedagogical approaches articulated in other chapters in this collection. But as higher education throughout the nation begins to make the shift from wired to mixed networking environments, I hope that other departments can use our story and key lessons (given below) to stabilize their own computing ground.

- Work closely and co-operatively with local ITS departments.
- Identify institutional spending priorities and patterns, learn to work with them, and/or find a way to re-articulate priorities and patterns.
- Identify the current figure of computing on your campus and work with that figure rhetorically, but always pay attention to the environment or ground of computing within your department.
- Seek funding from multiple sources, particularly sources that carry institutional weight.
- Explain technology needs not in terms of overcoming a deficiency but in terms of what an enhanced technological environment can contribute to the quality of undergraduate and graduate education on your campus.

These points have been central to our success in leveraging funds that can initiate and sustain a rich computing environment for GTAs. Once wireless recedes into the ground of higher education computing, if new institutional spending patterns are not established, and if English departments have missed the moment to perform an act of institutional critique and real change, I suspect that GTAs will have to provide their own equipment for teaching. The wireless revolution will not have resulted in the material changes that most graduate programs in English need.

ENDNOTES

1. The mid-semester questions were:
   1. How do you use your laptop and its wireless connectivity (i.e., a descriptive paragraph or two)?
   3. What benefits you have seen/felt/intuited?
   3. What drawbacks you have experienced?
   4. Is your computer (and its wireless connectivity) changing you in any subtle (or not so subtle) ways? Have you pushed aside old habits or friends because you have a laptop? Do you find yourself writing more email, doing more IM, doing more surfing, or other such things? For those who have been teaching at NDSU for more than a semester, is it changing your teaching?

   The end of semester 30-question survey is available at http://www.ndsu.nodak.edu/ ndsu/kbrooks/research/wireless.html

REFERENCES

Anderson, Daniel, Brown, Robin Seaton, Taylor, Todd, & Wymer, Kathryn. (2002). Integrating laptops into campus learning: Theoretical, administrative and instructional fields of play. *Kairos, 7*(1). Retrieved January 21, 2006, from http://english.ttu.edu/kairos/7.1/binder.html?response/CCI/index.html

Anson, Chris. (1999). Teaching writing in a culture of technology. *College English, 61*, 261-280.

Dean, Christopher, Hochman, Will, Hood, Carra, & McEachern, Robert. (2004). Fashioning the emperor's new clothes: Emerging pedagogy and practices of turning wireless laptops into classroom literacy stations. *Kairos, 9*(1). Retrieved January 21, 2006, from http://english.ttu.edu/kairos/9.1/binder2.html?coverweb/hochman_et_al/intro.html

Duffelmeyer, Barb Blakely. (2003). Learning to learn: New TA preparation in computer pedagogy. *Computers and Composition, 20,* 295-311.

Hernandez, Joshua. (2005, April). The gift of fire: The introduction of laptop computers into the English graduate program at NDSU. Paper presented at the Great Plains Alliance for Computers and Writing Conference, Mankato, MN.

Meeks, Melissa Graham. (2004). Wireless laptop classrooms: Sketching social and material spaces. *Kairos, 9*(1). Retrieved January 21, 2006, from http://english.ttu.edu/kairos/9.1/binder2.html?coverweb/meeks/index.html

McLuhan, Marshall, & McLuhan, Eric. (1988). *Laws of media: A new science.* Toronto: University of Toronto Press.

Porter, James E., Sullivan, Patricia, Blythe, Stuart, Grabill, Jeffrey T., & Miles, Libby. (2000). Institutional critique: A rhetorical methodology for change. *College Composition and Communication, 51,* 610-642.

Roen, Duane, Pantoja, Veronica, Yena, Lauren, Miller, Susan K., & Waggoner, Eric. (Eds.). (2002). *Strategies for teaching first-year composition.* Champaign, IL: NCTE.

Selfe, Cynthia L. (1999). *Technology and literacy in the twenty-first century: The importance of paying attention.* Carbondale: Southern Illinois University Press.

Selfe, Richard J. (2005). *Sustainable computer environments: Cultures of support in English studies and language arts.* Cresskill, NJ: Hampton Press.

8

A PROFILE OF STUDENTS USING WIRELESS TECHNOLOGIES IN A FIRST-YEAR LEARNING COMMUNITY

Loel Kim

Emily A. Thrush

Susan L. Popham

Joseph G. Jones

Donna J. Daulton

Recent developments in wireless technology offer students mobile, unconstrained access to the wealth of resources on the Web: Research, supplemental information, and lessons for their classes can be accessed and used in the classroom or anywhere on campus. Communicating with teachers and other students on e-mail, discussion boards, and blogs is possible 24 hours a day, seven days a week. Never before have multimodal capabilities been more widely available to and usable by so many students (Adewunmi, Rosenberg, Sun-Basorun, & Koo, 2003). In addition to technological developments, recent adoption of learning communities promises some students academic success by putting them into groups that seem to foster both social and intellectual interactions with faculty and each other, improving learning through collaboration and dialogue and increasing familiarization with the university culture (Gillespie, 2001; Lipson, 2002; Zhao & Kuh, 2004).

Because of the continued integration of wireless technology at the university and because its impact on student work, play, and communication are as yet unknown—particularly for those in learning communities—the goal for this chapter's study is to describe students' lives when equipped with laptops and wireless access. How do students use the technology to study and build ties to their learning community? To what extent and with whom? Which applications do they use and what patterns of use do they exhibit? When and where do they use their laptops most? This study examines technology use by 17 first-year students in a healthcare learning community as they navigate through their first semester at the University of Memphis (UM). From daily reports, questionnaire survey, and follow-up interviews, the study develops a profile of student technology use with wireless access and other common technologies. In particular, researchers wanted to see how students use technology as they learn to be college students, engaging in composition curricula and other first-year coursework, and as they develop and maintain academic and social ties. The observed patterns and frequency of use point to potentially fruitful areas of exploration as we develop pedagogies and formulate technology policies for the wireless campus.

BACKGROUND

Most universities are committed to improving student access to technology both to raise overall technology literacy rates and to ensure that their students benefit from any added educational value learning and working online may offer. Because first-year composition courses typically enroll a large proportion of incoming students across all disciplines each academic year, it makes good sense to target these courses for technology initiatives. But we know that users in organizations adopt technologies only if they fit systems of need, practice, and use already in place in their work lives, and reject forced use that does not benefit them (Grudin, 1990). Similarly, in order to understand the ways in which technology best supports educational goals, Scott Johnson, Elizabeth Gatz, and Don Hicks (1997) remind us that technology transfer, including both the development and spread of technology, needs further study. Thus, learning the ways in which students use technology can inform the development of truly effective curricula and policies.

LEARNING COMMUNITIES

Learning communities place students of similar interests into a cohort that progresses through a program of study together. Advocates believe the learning community structure helps students form strong bonds of friendship, collaboration and interactive learning (Jaffee, 2004; Schoem, 2002; Weissman & Boning, 2003). According to David Schoem (2002), the rationale behind a learning community assumes that "course content, pedagogy and learning are inherently intertwined," and that by forming "a community of scholars—both faculty and students—com[e] together for deeper learning" (p. 18). David Jaffee (2004) notes students in learning communities are more "likely to develop a deeper understanding of the material when concepts, topics, and debates introduced in one course are reintroduced and reinforced in another" (p. 1). Students claim that smaller class sizes make it more likely that they will interact with the teachers of these classes, and the continuity of the class cohort helps them to make connections among courses, resulting in more meaningful and relevant curricula (Weissman & Boning, 2003). Finally, students are also less likely to drop out of school when they engage socially and intellectually, when they feel like they "belong" (Tinto, 1998).

WIRELESS TECHNOLOGY

Increasingly, universities are expecting students to have laptops, whether purchased, rented, or—rarely—provided for them. However, reports of such initiatives lack evidence that these schools are restructuring their curricula or spaces to adapt to changes in the laptop learning environments. In a rapidly technologized world, the underlying assumption about computers in general and laptops, in particular, is that they will help students to learn better and work faster on existing assignments. However, little consideration has been given to accommodating individual preferences in study and work habits, including time of day, environment, comfort, lighting, and so on—features that may affect the ways in which technology supports student learning. Equally little research has attended to the ways in which wireless campuses and laptops might change interrelationships between students and faculty when the physical restrictions of separate buildings, classrooms, and departments change or even vanish. Perhaps wireless' greatest potential advantage for learning communities will be in shaping interaction among students—much as cell phones have increased our ability to work and communicate with each other on the go—or in creating a sense of virtual community through a changed sense of access to each other (Gant & Kiesler, 2001; Gay, Stefanone, Grace-Martin, & Hembrooke, 2001; Koo, Adewunmi, Lee, Lee, & Rosenberg, 2003; Swan, 2002).

An early study of students and laptops found that students' laptop use gradually caused dramatic changes in the structure of the school, including the classroom setting, length of class periods, and the division of curricula into subject areas (Spender, 1995, p. 112). Students seemed to prefer using their laptops sitting or lying on carpeted floors instead of at desks. The unorthodox setup allowed students to spread out materials and shift position frequently, avoiding back and neck strain. Additionally, the typical 40-50 minute class period was found too short to allow students to engage with projects, an activity that laptops in the classroom seemed to promote (Spender, 1995, p. 112). Finally, instead of discrete assignments, laptops seemed conducive to more integrative student projects that spanned courses (Spender, 1995, p. 113).

In another study, first-year students at a small, private college reported attitudes towards laptop use that showed, overall, a strong negative correlation between laptop use and a sense of a learning community (Demb, Erickson, & Hawkins-Wilding, 2004). However, students who lived off campus reported that they more frequently interacted in "activities related to the creation of an intellectual community . . . 'a community of learners'" (Demb et al., 2004, p. 397). Researchers concluded that when the laptop was the primary computer for the student, it was used much more than when it was a secondary computer. Also, students who used the laptops as their primary computer used them more for group projects, communication with their advisors, and engagement in chatrooms, and they felt laptops contributed to a stronger sense of community (Demb et al., 2004, p. 397). Thus, wireless laptops may contribute to a growing sense of community when physical access is limited.

Research examining community building through technology focuses primarily on online courses, and not those that mix technology with weekly face-to-face contact (Brown-Yoder, 2003; Dial-Driver & Sesso, 2000). Although some studies have shown that students can benefit from more frequent, individualized, and immediate contact with faculty through online access (Demb et al. 2004; Jaffee, 2004; Kiesler & Sproull, 1987; Schofield & Davidson, 2003), some studies raise questions about problems of online access (Smelser, 2002; Yena & Waggoner, 2003). Lynne Smelser (2002) argues that although some students may find increased opportunities for academic engagement, other students, especially those who wrestle with the technology or struggle to find an academic voice, may struggle even more conspicuously in an online environment. Another study reported that when students engage in frequent e-mail contact with the instructor, students find the communication less personal than face-to-face interaction (Vonderwell, 2003). This corroborated other studies in which the instructor provided quick and frequent feedback at the beginning of the semester, but as the term progressed, both quality and quantity of feedback decreased. Finally, student perceptions of online teacher feedback show mixed results—the increased capacity of "media-rich" technology alone is not sufficient to ensure improved communication between students and teachers. The quality of teacher feedback is ultimately dependent on individual differences (Kim, 2004).

This current study focuses on wireless activity outside the classroom, because so many forms of technology are now common to undergraduate students as they work and communicate in college, and we simply do not know how they are using them. Although we expect wireless connectivity to affect curricula, it may also reshape the physical configurations of campuses. Although many studies on use of wireless campus networks focus on technical issues (primarily to determine where to invest institutional resources), some studies have shed light on how wirelessness is likely to impact the student experience. For example, David Schwab and Rick Bunt (2004) found that students were more likely to use the wireless network in communal areas such as a student center, library study areas, and spaces adjacent to computer labs (p. 7). Conversely, David Kotz and Kobby Essien's 2002 study found that students used wireless laptops most in their dormitories (p. 6). Similarly, Charles Crook and David Barrowcliff (2001) identified studying as a sedentary activity, a position incompatible with predictions for ubiquitous computing found elsewhere in the research. They argue, as do Kotz and Essien (2002), that student computing would be located primarily at desktops in the dorm rooms.

Aside from place of use, the study in this chapter was designed to identify network and application use as well, to provide insight into how wireless technology, rather than simply fitting into current educational practices and facilities, might shape the ways in which educational institutions will function in the future.

STUDY DESCRIPTION

We wanted to answer the general question: In what ways will students in an integrative learning community use wireless, laptop computers to work, play, and communicate? To answer it, 17 learning community students in the final six weeks of the fall semester were equipped with notebook computers with wireless capability to use as required in their composition class and as they otherwise wished. Students were signaled three times per day and entered in technology logs their technology use and interactions with people.

The students' first-year composition course instructor agreed to participate in the study by allowing the research group to observe his class and incorporating technology into his curriculum. In return, he was given a laptop to use, some sample assignments incorporating technology, access to technology workshops, and the opportunity to register his class for an online course management system.

Participants

Seventeen first-year undergraduate students participating in a pre-med learning community at the UM were recruited. The community comprised students who had expressed interests in medicine, veterinary science, dentistry, ophthalmology, nursing, and physical or occupational therapy. The student cohort stayed together through three courses over two semesters of the required first-year writing courses; the study took place in the first semester. Students were given the use of a laptop computer for six weeks of the first semester and the entire second semester, and they received candy bars and a small cash incentive for their participation in follow-up interviews as well.

Twelve (71%) students are female and five (29%) are male. They are all 18 years old, and self-reported their races as: 4 black, 9 white, 3 Asian, and 1 mixed. Students represented moderately well-educated families: All parents had completed high school, a majority of parents (61.8%) had at least some college, and one-third (32.4%) had completed college degrees.

Materials

Hardware. Each student was issued a Dell Latitude D600 with an Intel Pentium 1.5 GHz processor with 512 MB of RAM, and Microsoft WINDOWS XP PROFESSIONAL operating system. In addition, each laptop had a wireless internet card and DVD/CDRW. Technical support was available through campus services. Each student also had a cell phone which was used to contact him or her throughout the study.

Software. Computers were originally configured with Microsoft OFFICE XP PROFESSIONAL with FRONT PAGE, Netscape 7.1, INTERNET EXPLORER,

SYSTAR10.2, Roxio EASY CD CREATOR 5, Symantec CLIENT SECURITY, the SAS SYSTEM RESPONDUS, QVTNET, QUICKTIME, SPSS FOR WINDOWS, and Adobe ACROBAT 6.0.

Data Gathering

The profile of student technology use was developed from a number of sources:

- **technology logs** (Appendix A). Students were contacted on cell phone three different times during each of 47 consecutive days by a research assistant using a professional web-based calling service. Calls times were random within each of the three time periods: 12 midnight through 8:00 a.m., 8:00 a.m. to 4:00 p.m., and 5:00 p.m. to 12:00 p.m. (we oversampled calls during the midnight to 8 a.m. period, when we expected response rates would run low). If using technology when notified, students selected from a list of technology types to indicate their active, primary use as well as secondary use. In addition, they were asked to indicate the programs they were using and the people, if any, they were in contact with.
- **demographic and technology questionnaires** (Appendices B and D). Students were asked to supply personal information relevant to the study including age, race, parents' education levels, and the length of time they had access to computers. This item was designed to determine individual length of exposure to and direct experience with a computer. All participants but one came from families with computers in their homes, with an average of 10.9 years as the age when their family acquired one. A high proportion of students also had their own computers (70.6%), starting at an average age of 16.1, and 91% of them indicated that they felt confident using a computer. Further, students were familiar using computers to write, reporting that they were expected to use word processing programs in high school.
- **follow-up questionnaire** (Appendix C). Poststudy questionnaires helped us flesh out data from the logs and included items probing: the extent that students customized their laptops, self-reported frequency of use, technology problems, and locations they frequented to use the technology.

- **follow-up interviews.** The instructor and some students added insights to their experience with the laptops, which were tape recorded and transcribed. Their comments further illustrated their responses to the questionnaires and in the logs.

RESULTS AND DISCUSSION

Throughout the six weeks of the study, students' technology logs were collected weekly.[1] At the end of the study, responses were compiled and students filled out exit questionnaires. Some students were interviewed for more detail. Our findings are presented in this section.

How Frequently and When Were Students Using Technology?

Responses to the first question from the log give an overall sense of students' frequency of technology use throughout the day. Table 8.1 shows that students were using technology 17.7% of the time:

How should we interpret this figure? First, we expected that students would be inactive during certain times of the day—when they slept and, perhaps, during meals—approximately 11 hours (8 hours sleep and 3 hours for meals) of probable "no" responses to the logs, or 45.8% of the total possible time students have available in the day. On the other hand, the 17.7% rate converts to 4.2 hours per day using technology.[2] When this total tech-

TABLE 8.1. AT THE MOMENT YOU WERE CALLED/BEEPED,
WERE YOU USING ANY TECHNOLOGY?

	FREQUENCY	PERCENT OF RESPONSES
YES	207	17.7
NO	964	82.3
TOTAL ACTUAL	1,171	100
TOTAL POSSIBLE	2,397	
RESPONSE RATE	48.9%	

nology use figure is broken down for laptop use, the figure reduces to 6.7%, or 1.62 hours per day, 11.34 hours per week. This moderate rate of use was supported by students' self-report in the post-interviews (3.3/5). To put this in a larger context, a 1999 study of 651 freshmen at the University of Colorado at Boulder showed students spent 6.7 hours per week working on computer ("Undergraduate Students–1999," accessed online). Given total access to their laptops during the study, our students' usage—almost twice that of the Boulder survey—seem in line. Total increased usage suggests that in this first semester in college learning community students are making frequent, if not heavy use of technology available.

PATTERNS OF USAGE

In addition to how much time students spent using technology, their patterns of use illuminate student preferences and behaviors. Our data showed heavier usage during certain times of day more than others, as shown in Figure 8.1. Heaviest usage occurred in the mornings between 8:00 a.m. and 9:00 a.m., with the second period occurring between 6:00 p.m. and 7:00 p.m. Usage spiked at 2:00 p.m. and 9:00 p.m., and there was a modestly steady rate throughout the night, from 2:00 a.m. to 7:00 a.m.

Figure 8.1. At What Times of Day Were Students Most Often Using Technology?

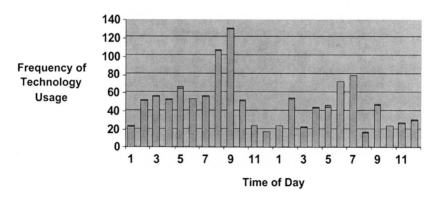

These periods of technology use seem reasonable and what we might expect. In particular, the morning spike occurs around waking hours and prior to students' writing class, which took place three days per week from 10:20 a.m. to 11:15 a.m. This suggests that students prepared for class just prior to class (a behavior long suspected by teachers).

How Mobile Was Student Technology Use?

Our group of students reported they used their laptops most frequently, first, in class (55%), followed by in their dorms (36.4%), and third, at other places on campus (9.1%). Frequency of use seems to be related to place of use. Students who rated themselves highest in frequency of use (4 and 5) tended to select places other than the classroom as their most frequent place of use. One of them, Hannah, lived off campus and found the laptop necessary to get her schoolwork done. She also enjoyed the collaborative work style it allowed:

> [The laptop was] convenient . . . because . . . we could go to Starbucks and type our paper together. Regan and I went to Perkins and typed our papers together and we would sit in my living room and type our papers together and then each get on the internet and send it to UM Drive or something.

Another student observed the convenience wireless allowed her:

> Before I became an RA in [a dorm] that had a computer lab . . . it was so hard trying to getting on computer. I would wait until the last minute to go to the computer and I wouldn't be able get on. With the laptop in my room . . . [it's] addicting . . . I didn't have to get up and go any-where . . . I would be able to get on and save documents for my class-es. I can print it out wherever.

On the other hand, the three female students who reported the lowest rate of use (1 and 2) reported their primary place of use was the classroom. Most students seemed to carry their laptops only when they knew they needed them in class, and students who reported themselves to be low users used the laptops predominantly in the classroom. Particularly for these students, more structured in-class computer use might be effective in helping them to gain skills that savvier technology users seem to gain on their own. Also, although today's laptops weigh only a fraction of the weight of early portable computers, size and weight could be deterrents to mobile computer use. As Katie observed:

> Umm, [I didn't] . . . take it for class all the time . . . like when [I] did-n't need it, considering all the other books I had. It made it a whole lot heavier on my shoulders. Yeah, I only take it when I know I absolutely need it.

Additionally, poorly incorporated wireless technology in the classroom offers students more opportunity to *disengage* with the classroom activity. As one student, Abdul, reported of one class meeting, "It was really bor-ing—most of us pulled out our laptops and we were just sitting there IM'ing someone . . ." However, the instructor's observation about problems with wireless in the classroom suggests, perhaps, a more serious concern: Effectively engaging in a classroom assignment handled through wireless technology requires the ability to already communicate well:

> Just because you have a computer—if you can't articulate and write, it doesn't matter, anyway. You can go on the Internet all day long and search and do all kinds of things and learn stuff, but then you type it out and you have subject and verb disagreements. I noticed . . . the ones that couldn't write were the ones most of the time that were goof-ing off, looking at the internet, looking up Wal-Mart—what they had on sale.

In addition, for inexperienced instructors, establishing protocols so that stu-dents and the instructor were "on the same page" during class would help to manage interactions. Our instructor noted afterwards:

> Students were told: "This is a technology class. Here is your laptop. Bring it to class everyday." So their take on it was—it's always open. If I'm talking or we're discussing things they always have it open, regard-less—to do what they want to do. It probably would have been up to me to say, "Ok, shut your computers. We're gonna talk about this. And then open your computers."

In their study, Demb et al. (2004) found that 52 % of students felt faculty did not incorporate technology into their teaching very well (p. 391). Researchers further observed that moving technology use in the classroom beyond the first wave of early adopters requires focused teacher training and support, as the population of faculty differs in technology skills and motivation from early adopters (pp. 396-397). As noted earlier, the instructor in our study expressed frustration with the challenge and amount of effort it took to incorporate technology into his course curriculum. Even though disappointing, this finding points to a primary problem in adapting technology effectively into the classroom: Often instructors of composition classes are pressed for time by heavy course loads, and high enrollments. Additionally, if, as is common for first-year composition programs, the instructor is a TA, then we are relying on the least experienced teachers to develop new pedagogies. Other studies have noted teacher training as essential to effective technology adoption in classrooms (Kimme Hea, 2002; Palmquist, Kiefer, Hartvigsen, & Goodlew, 1998; Selfe, 1992). At least in this first generation of wireless technology use in the classroom, new teachers—who may model their teaching practices on their experiences as students—may simply be overly burdened for successful technology adoption. This is particularly unfortunate since most students in our study expressed the desire to learn technology in their classes:

> I would probably try and umm, teach them something on the laptop instead of just giving someone a laptop. I wanted to learn how to use WEB-CT. I didn't get to use it. I would be like "Here this is what you can do with the laptop," you know. Make it kind of fun.

PROBLEMS WITH THE TECHNOLOGY

Overall, students reported moderate frequency of problems with the laptops (3.5/5), and students who reported high problem rates tended to be low users. Although the most frequently reported problem was network access (9/16 or 56.3%), students also reported problems commonly encountered by users. For example, James, a high user, noted:

> When I got back [after the winter break] someone else got my laptop, and I got stuck with theirs and it was loaded with spyware and junk. The whole computer was messed up. So, I ran a spyware program on it and found 500 and something infected files! Every time I'd get on, it would say something like, "Online casino" or something stupid. And then there's like Hello Kitty screen savers on there. I tried to go through and delete all those. I couldn't even get rid of all those. It was like a cockroach—it wouldn't die.

What Kinds of Technology Were Students Using?

The second question in the log probed the type of technology students used. As shown in Table 8.2, when students were contacted, the primary technologies they were using were, first, their laptops (40.5%), followed by DVD or TV (26.2%), and then cell phones (19%).

The third question anticipated that students would often have a number of technologies open, some being used secondarily or passively. Of those students who responded that they did have another technology on in the background, the most common other item was, by far, the DVD/TV (32.6%), followed by cell phones (19.4%), CD/MP3/iPod (17.4%), and laptops (16.7%).

These two questions provide us with a picture of the core of technologies students use. Students tend to use the laptop most frequently, and all students reported having more than one technology open at a time, in particular, the DVD/TV. Thus, these students regularly work in environments where multiple sources of information may be competing for their attention, or where they have become attuned to working with technological "white noise." Student preference for multitasking and background noise might indicate that optimal structuring of wireless class/study spaces will differ from current assumptions underlying environmental design. Whether or not future classrooms, dorms, libraries, and study spaces need to be reconfigured as far as seating, work spaces, and managing levels of noise are concerned, are interesting questions educators currently face.

One notable observation is that although 70.6% of our students reported owning desktop computers, use of those desktops fell well near the bottom of the list. Aside from the possible novelty of the wireless laptops, this usage indicates students switched primary use to the laptops, and it corresponds with observations *Internet.com* and many other market research and news sources have made about the popularity of laptops, whose sales have been increasing as desktop sales decline (Greenspan, 2003).

TABLE 8.2. TECHNOLOGY USE

Type of Technology	ACTIVE USE		SECONDARY USE		
	Frequency	% of Responses	Frequency	% of Responses	% of Possible Responses
Laptop	79	40.5	24	16.7	2.0
Desktop	11	5.6	6	4.2	.5
Cell phone	37	19.0	28	19.4	2.4
Land line telephone			3	2.1	.3
PDA			2	1.4	.2
DVD (laptop)	2	1.0	4	2.8	.3
DVD (TV)	51	26.2	47	32.6	4.0
CD/mp3/iPod	2	1.0	25	17.4	2.1
Digital video/camera	7	3.6	2	1.4	.2
Calculator	2	1.0	1	.7	.1
FAX/copier/scanner/media reader	4	2.1	2	1.4	.2
Total Responses	195	100	144	100.1	12.3
TOTAL Possible Responses	1,171		1,171		

Which Applications Were Students Using?

Question 4 asked students the type of software applications they were actively using (see Table 8.3). Word processing (47.9%) and Web browsers (31.3%) ran significantly ahead of other choices, with email (9.4%) falling a distant third place. Unlike students in the early days of computers and writing, all of our students connect computer use with writing well (Appendix D).

We imagined students might vary in their use of the technologies depending on whether they were responding on a weekday or weekend. On the weekend did they, for example, use the laptop less and use cell phones or DVD/TVs more? To examine this more closely, we compared students' Yes and No responses to Question 1 in the Technology Log and computed a percentage of technology use for each day, summarizing across all subjects for each date. Table 8.4 summarizes this analysis:

Surprisingly, we found no significant differences in the proportions of technology use on weekdays (18.58%) versus weekends (15.63%) when tested by Chi Square (1) = 1.4776, p = .224. Student levels of technology use were similar across weekdays and weekends.

TABLE 8.3. WHICH APPLICATIONS WERE STUDENTS ACTIVELY USING?

TYPE OF APPLICATION	FREQUENCY	% OF RESPONSES
MS Word	46	47.9
database manager	1	1.0
Netscape/Explorer	30	31.3
Dreamweaver	1	1.0
Photoshop	1	1.0
E-mail	9	9.4
IM	1	1.0
Other	7	7.3
Total Responses	96	99.9
TOTAL Possible Responses	1,171	

TABLE 8.4. PERCENT OF PRIMARY TECHNOLOGY USE
FOR WEEKDAYS VERSUS WEEKENDS

		WEEKDAY	WEEKEND	TOTAL
YES	Frequency	152	55	207
	Percent	12.99	4.70	17.69
	Row Pct	73.43	26.57	
		18.58	15.63	
NO	Frequency	666	297	963
	Percent	56.92	25.38	82.31
	Row Pct	69.16	30.84	
		81.42	84.38	
Total	Frequency	818	352	1,170
		69.91	30.09	100.00

What Activities Were Students Engaged in and With Whom?

Question 5 characterized the nature of the activities students engaged in and how much time they allocated to them (see Table 8.5).

Not altogether unexpectedly, we found our students worked approximately as much as they played, and when they did relax, they did so much more often with others than alone. However, students did not engage as

TABLE 8.5. WHAT PATTERNS OF WORK AND PLAY
DID STUDENTS EXHIBIT?

TYPE OF ACTIVITY	FREQUENCY	% OF RESPONSES	% OF POSSIBLE RESPONSES
Working (school)	69	38.5	5.9
Working (job)	4	2.2	.3
Relaxing alone	31	17.3	2.6
Relaxing with others	73	40.8	6.2
Managing	2	1.1	.2
Total Responses	179	99.9	12.3
TOTAL Possible Responses	1,171		

much with their learning community friends as other people. Question 6 from the Technology Log asks whether or not students were communicating with other people and if so, with whom. In Table 8.6, a pattern of student interactions emerge: First, nonschool friends (43%) are by far the most frequent contact when students are using technology, with family (21.9%) and non-LC students (20.3%) placing close second and third places.

Contacts with learning community students (10.9%) ran behind the first place contacts. We suspect this contact would change over time. At this point, after all, they have only known their learning community classmates for less than one semester and have not had a chance to forge relationships as deep as those of their precollege friends. As for fostering more interaction with other learning community students, Francesca noted that the laptop may have had an isolating effect when she was around other students, "[it] probably kept me from talking to other people instead. I mean, yeah, it kept me from talking to people." But the choice of communication modality may differ depending on the type of communication the user engages in—work-related or social. Although communication among learning community classmates might have been fostered had the course included group projects, individual student preferences and situation-related circumstances complicate the picture. In one study of learning communities with online access, researchers found 83% of the students working in collaborative groups reported doing most of their communication with group members face-to-face, not online (Glaser & Poole, 1999).

TABLE 8.6. WHAT PATTERNS OF CONTACT WITH PEOPLE DID STUDENTS EXHIBIT?

PERSON	FREQUENCY	% OF RESPONSES	% OF POSSIBLE RESPONSES
LC students	14	10.9	1.2
Other students	26	20.3	2.2
Nonstudent friends	55	43.0	4.7
Family	28	21.9	2.4
Coworker	4	3.1	.3
Faculty	1	.8	.1
Nonfaculty UM personnel	0	0	0
Total Responses	128	100.0	12.3
TOTAL Possible Responses	1,171		

STUDENTS COMMUNICATING WITH FACULTY

Given the mixed, but overall encouraging prior research on teacher-student communication through e-mail, and perhaps our own hopefulness as teachers, we had expected students would report frequent contact with faculty, but unfortunately, this was not the case. Students reported almost no communication with faculty, and when we asked the course instructor how many of the students e-mailed him, he estimated, "Umm, I would say . . . 60% maybe," but added that the figure was over the entire semester and that communication focused on course management topics, such as absences and late assignments. Similarly, in a study of technology use for teaching, John Savery (2002) found more than 90% of the faculty reported using e-mail to communicate with students for instructional purposes, much higher than any other reported use of technology (p. 7). However, only 38% of the students reported that e-mail had been used for instructional purposes more than five times during the semester. Savery explained the finding by noting that a teacher might send and manage e-mails from many students, whereas students are exchanging many fewer e-mails with just one teacher. In addition, although broadcasting messages to a class through e-mail may serve an important function for the teacher in managing a class, student perceptions of instruction may mean more individual e-mail feedback about lessons, assignments, or individual progress.

RETENTION AND CREATING FAMILIAR ENVIRONMENTS

With 15 of the original 17 students returning to the learning community in the spring semester, this group exhibited a similarly high retention rate as previously observed (P. Krech, personal correspondence, Feb. 16, 2005). Returning students enthusiastically welcomed the return of their laptops, which had been collected over winter break. Students seemed especially keen to get the same one they used in the fall semester, checking serial numbers and desktop configurations to ensure they did. This behavior suggests, not surprisingly, that the effort that went into personalizing laptops and the desire to return to the familiar environment are important factors in technology being useful to students. Comments like, "I was familiar with what was and wasn't on my computer" and "some of my personal files, such as music, pictures, and documents were all saved on there," occurred frequently in postinterviews.

In fact, students noted that they would have probably customized their laptops more extensively if they actually owned them. Hannah noted her restrained downloading: "If it had been mine and I was going to have it for years and years and not have to give it back I would do [more] stuff to it." Likewise, Francesca did not feel completely free with the laptop, "I really didn't try to explore anything because I knew it was your computer. I didn't want to break anything." In addition, Hannah expressed a sense of privacy, "'cause you put all your stuff in there that I don't want everybody in the world to see, my pictures and stuff." Some of the other changes students made:

> [I] rearranged desktop, downloaded games, saved internet sites
> I added limewire, music, weatherbug, and AIM to it
> I added some programs that I originally had on my desktop to make it familiar.
> Games! We downloaded a few games.

These changes ranged from those affecting the look and location of items on their desktops, such as changing their wallpaper or screensaver, to changes affecting the content or functionality of their laptops, such as downloading programs, games, and music. One-third of students reported downloading instant messaging software.

CONCLUSION

This study indicates that the benefits of wireless access for learning communities are mixed. Although our students used a variety of technologies, their use, as measured both by the technology logs and by self-report, seemed to be predicated most heavily by requirements and expectations of the classroom. This is not necessarily bad—thoughtfully developed curricula incorporating technology may have more of an effect on higher technology literacy than other factors, including ownership. We saw some indication that even when students are from middle-class families with technology present in their homes, some students, particularly females, may be less inclined to use technology beyond the basics. This may be useful to know when planning wireless implementation where low-use students form a substantial portion of the student population.

Whether or not wireless access has a positive effect on learning community students' social ties is also unclear. However, many of our observations in this short study indicate that longitudinal studies are necessary to complete the picture of student communication in learning communities. After all, friendships grow over time. Our hunch is that a study in which observations could be made past the first semester would show increasing interaction among learning community students. In addition, further study of wireless supported pedagogies as well as online student-teacher communication may shed light on how technology may be best used to enhance student interaction with teachers.

In addition, we should not overlook the presence of other technologies, whose use, particularly DVD/TV, cell phones, and CD/MP3/iPod, comprise a strong presence in student environments. The ways in which these technologies might affect student learning and community-building merit further study. Environmental studies of classroom, dorm, and other campus space design can add to our understanding of how wireless access is changing the spaces in which we teach and students learn.

ACKNOWLEDGMENTS

The authors wish to thank the Advanced Learning Center at The University of Memphis for financial and technical support through the Technology Access Fee Innovation Grant. Special thanks to George Relyea, Research Assistant Professor, Center for Community Health, The University of Memphis, for his contributions to research design and analysis. Finally, our thanks to the book's reviewers, whose insights helped us improve the chapter.

APPENDIX A
TECHNOLOGY LOG

At the moment you were called/beeped, were you using any technology?
❏ YES ❏ NO

If NO, then you are done with this entry and you may STOP HERE.

If YES, then PLEASE CONTINUE:

Check off below all applicable categories that describe your technology use at the time you were beeped.

Kind of technology you were *actively* using
___ Laptop   ___ PDA   ___ digital video/camera
___ Desktop   ___ dvd/TV   ___ cd/mp3/iPod
___ cell phone   ___ dvd (laptop)   ___ calculator
___ Land line   ___ FAX, copier, scanner, media reader

Kind of technology you were *passively* using
___ Laptop   ___ PDA   ___ digital video/camera
___ Desktop   ___ dvd/TV   ___ cd/mp3/iPod
___ cell phone   ___ dvd (laptop)   ___ calculator
___ Land line   ___ FAX, copier, scanner, media reader

Kind of application you were *actively* using
___ MS Word   ___ Netscape/IExplorer   ___ a statistical application
___ MS Excel   ___ Dreamweaver   ___ e-mail   ___ IM
___ database manager   ___ Photoshop   ___ Other

Primary type of activity you were engaged in
___ working (school-related)   ___ relaxing alone   ___ Managing
___ working (job-related)   ___ relaxing with others

People you were in contact with
___ learning community students   ___ family   ___ faculty
___ other students   ___ co-worker   ___ non-faculty UM staff
___ non-student friends   ___ no one

APPENDIX B
DEMOGRAPHIC QUESTIONNAIRE

Year born: _____   Gender: ___ F   ___ M

Check the categories below that best describe your racial and ethnic identity:

___ Black or African-American   ___ Native Hawaiian or Other Pacific Islander

___ Hispanic or Latino   ___ American Indian or Alaska Native

___ White   ___ Arab-American or Middle Eastern

___ Asian   ___ Other _____

Year of Study: ___ 1st   ___ 2nd   ___ 3rd   ___ 4th   ___ 5th

Major:_____

Minor:_____

If undecided above, area of interest:_____

Age you were when your family had a first computer: _____

Age you were when you had your "own" computer: _____

Check off your parents' years of formal education:

Number of years finished

Mother	Father
___ 1-6	___ 1-6
___ 7-8	___ 7-8
___ 9-10	___ 9-10
___ 11-12	___ 11-12
___ some college	___ some college
___ finished college	___ finished college
___ some graduate courses	___ some graduate courses
___ graduate degree (M.A., M.S., M.B.A., for example)	___ graduate degree (M.A., M.S., M.B.A., for example)
___ Ph.D.	___ Ph.D.

APPENDIX C
LAPTOP PREFERENCES QUESTIONNAIRE

When the laptops were reassigned to you this semester, did you try to get the same laptop you used last fall?

❏ YES ❏ NO

If YES, why did you want the same laptop back?_____

In what important ways did you change the laptop from its original state? Please include *any* types of changes you made to the arrangement or content of the laptop that made it more useful or pleasant for you to use.

Please check off any types of software you downloaded to your laptop:

___ updates of software already installed when you received it

___ software/drivers for peripheral equipment

___ software for school-related work or activities

___ software for recreation

___ other (please list)

Please check off any other downloads to your laptop:

___ music   ___ games

___ images (photos, clip art, Web art)   ___ pdf files

___ other (please list)_____

If you downloaded any software or files to your computer, where did you find out about the downloads? Please check all that apply.

___ learning community classmates   ___ teacher   ___ other classmates

___ friends   ___ family members   ___ colleagues at work

___ automatic update notice   ___ other (please list) _____

How frequently did you encounter problems when using your laptop?

Extremely high		Moderate		Rarely
1	2	3	4	5

If you indicated you experienced problems in the above item, please check off the area below that caused you the most trouble.

___ hardware (the laptop itself, ports/plug-ins, hard disk, memory or storage problems)

___ network (ability to connect to the Internet smoothly and dependably, speed)

___ applications (software programs)

___ other (please describe) _____

If you could customize the laptop in any way you wanted, what would you do?_____

Overall, how frequently do you use your laptop?

less than once per week	a few times per week	several times per week	once or twice per day	several times per day
1	2	3	4	5

Where did you most often use your laptop? Please pick the **top three** locations and number them in order of frequency, 1 = most:

___ dorm room, own room at apartment or home

___ common living spaces (living room, dining room, den, etc.) at apartment or home

___ library

___ classroom

___ other campus location (Tiger Den, Univ Ctr lobby, etc., please describe) _____

___ other off-campus location (Starbucks, restaurant, etc., please describe)

_____

APPENDIX D
WRITING AND TECHNOLOGY QUESTIONNAIRE

Directions: Below are a series of statements about writing and technology. There are no right or wrong answers to these statements. Please indicate the degree to which each statement applies to you by circling whether you (1) strongly agree, (2) agree, (3) are uncertain, (4) disagree, or (5) strongly disagree with the statement. Thank you for your cooperation.

1 = strongly agree   2 = agree   3 = are uncertain   4 = disagree   5 = strongly disagree

1. I always use a computer when I write compositions.	1	2	3	4	5
2. I prefer to write several versions of a composition before I hand it in for evaluation.	1	2	3	4	5
3. I always cut and paste to reorganize my compositions when I write on the computer.	1	2	3	4	5
4. My high school emphasized computer skills for writing.	1	2	3	4	5
5. I use the Internet for all my research.	1	2	3	4	5
6. I feel confident using the computer when I write.	1	2	3	4	5
7. I usually e-mail versions of a composition to friends before I turn it in for evaluation.	1	2	3	4	5
8. I can't imagine writing without using a computer.	1	2	3	4	5
9. I rely on the computer to check my spelling and grammar.	1	2	3	4	5
10. I was expected to use a computer for my writing assignments in high school.	1	2	3	4	5
11. I received little or no instruction in using technology for my writing assignments in high school.	1	2	3	4	5
12. I prefer to discuss versions of a composition with others before I turn it in for evaluation.	1	2	3	4	5
13. Using a computer makes my writing better.	1	2	3	4	5

ENDNOTES

1. Students were contacted around the clock for an extended period of time, a fairly heavy burden on them. As might be expected, response rates varied widely by student, but the overall rate of response was acceptable at 48.9%.
2. This average, of course, includes those students who reported very low use and those who might be using their laptops or other technology for several hours each day. It also glosses patterns of use in which students might not use technology at all on some days and use it heavily on others.

REFERENCES

Adewunmi, Adegbile, Rosenberg, Catherine, Sun-Basorun, Adeoluwa, & Koo, Simon G. M. (2003). Enhancing the in-classroom teaching/learning experience using wireless technology. *Proceedings IEEE Frontiers in Education Conference, 3*, 1-6.

Brown-Yoder, Maureen. (2003). Seven steps for successful online learning communities. *Learning and Leading with Technology, 30*(6), 14-21.

Crook, Charles, & Barrowcliff, David. (2001). Ubiquitous computing on campus: Patterns of engagement by university students. *International Journal of Human-Computer Interaction, 13*, 245-256.

Demb, Ada, Erickson, Darlene, & Hawkins-Wilding, Shane. (2004, December). The laptop alternative: Student reactions and strategic implications. *Computers & Education, 43*, 383-401. Retrieved October 9, 2005, from *Elsevier Science Direct Database* doi:10.1016/j.compedu.2003.08.008

Dial-Driver, Emily, & Sesso, Frank. (2000). Thinking outside the (classroom) box: The transition from traditional to on-line learning communities. Retrieved October 9, 2005, from *ERIC* database. Accession No: ED448457.

Gant, Diana, & Kiesler, Sara. (2001). Blurring the boundaries: Cell phones, mobility, and the line between work and personal life. In Barry Brown, Richard Harper, & Nicola Green (Eds.), *Wireless world* (pp. 121-131). New York: Springer-Verlag.

Gay, Geri, Stefanone, Michael, Grace-Martin, Michael, & Hembrooke, Helene. (2001). The effects of wireless computing in collaborative learning environments. *International Journal of Human-Computer Interaction, 13*, 257-276.

Gillespie, Kay Herr. (2001). Editor's page: The concept of learning communities. *Innovative Higher Education, 25*(3), 161-163.

Glaser, Rainer E., & Poole, Melissa J. (1999). Organic chemistry online: Building collaborative learning communities through electronic communication tools. *Journal of Chemical Education, 76*, 699-703.

Greenspan, Robyn, (2003, July 2). Notebooks overthrow the desktop. Retrieved November 2, 2004, from http://www.internetnews.com/bus-news/article.php/ 2230831

Grudin, Jonathon. (1990). Groupware and cooperative work: Problems and prospects. In Ronald M. Baecker (Ed.), *Groupware and computer-supported cooperative work* (pp. 97-105). San Mateo, CA: Morgan Kaufmann.

Jaffee, David. (2004). Learning communities can be cohesive and divisive. *The Chronicle of Higher Education, 50(44),* B16.

Johnson, Scott D., Gatz, Elizabeth F., & Hicks, Don. (1997). Expanding the content base of technology education: Technology transfer as a topic of study. *Journal of Technology Education, 8(2),* 35-49.

Kiesler, Sara, & Sproull, Lee. (1987). *Computing and change on campus.* New York: Cambridge University Press.

Kim, Loel. (2004, February). Online technologies for teaching writing: Students respond to teacher comments in voice and written modalities. *Research in the Teaching of English, 38,* 304-337.

Kimme Hea, Amy C. (2002). Rearticulating e-dentities in the web-based classroom: One technoresearcher's exploration of power and the World Wide Web. *Computers and Composition, 19,* 331-346.

Koo, Simon G. M., Adewumni, Adegbile, Lee, Jeongjoon, Lee, Yat Chung, & Rosenberg, Catherine. (2003). Graduate-undergraduate interaction in wireless applications research and development. *Proceedings—Frontiers in Education Conference, 3.*

Kotz, David, & Essien, Kobby. (2002). Characterizing usage of a campus-wide wireless network. Dartmouth computer science technical report TR2002-423. Retrieved October 9, 2005, from http://cmc.cs.dartmouth.edu/cmc/papers/ kotz:campus-tr.pdf

Lipson-Lawrence, R. (2002). A small circle of friends: Cohort groups as learning communities. *New Directions for Adult and Continuing Education 2002, 95,* 83.

Palmquist, Mike, Kiefer, Kate, Hartvigsen, James, & Goodlew, Barbara. (1998). *Transitions: Teaching writing in computer-supported and traditional classrooms.* Greenwich, CT: Ablex.

Savery, John R. (2002). Faculty and student perceptions of technology integration in teaching. *Journal of Interactive Online Learning, 1(2),* 1-16.

Schoem, David. (2002). Transforming undergraduate education: Moving beyond distinct undergraduate initiatives. *Change, 34(6),* 50-55. Retrieved January 27, 2006, from ERIC full-text database.

Schofield, Janet W., & Davidson, Ann Locke. (2003). The impact of Internet use on relationships between teachers and students. *Mind, Culture, and Activity, 10,* 62-79.

Schwab, David, & Rick Bunt. (2004). Characterising the use of a campus wireless network. *IEEE Info-Com.* Retrieved July 22, 2005, from http://www.ieee-info-com.org/2004/Papers/18_1.PDF

Selfe, Cynthia L. (1992). Preparing English teachers for the virtual age: The case for technology critics. In Gail E. Hawisher & Paul LeBlanc (Eds.), *Re-imagining computers and composition: Teaching and researching in the virtual age* (pp. 24-42). Portsmouth, NH: Boynton/Cook.

Smelser, Lynne. (2002). Making connections in our classrooms: Online and off. Retrieved July 22, 2005, from *ERIC* database. Accession No.: ED464323

Spender, Dale. (1995). *Nattering on the net: Women, power, and cyberspace.* North Melbourne, Vic: Spinifex Press.

Swan, Karen. (2002). Building learning communities in online courses: The importance of interaction. *Education, Communications and Information, 2,* 23-49.

Tinto, Vincent. (1998). Colleges as communities: Taking research on student persistence seriously. *The Review of Higher Education, 21,* 167-177.

University of Colorado. (1999). Undergraduate students—1999: How students spend their time. Retrieved October 9, 2005, from http://www.colorado.edu /pba/surveys/ ug/99/time.htm

Vonderwell, Selma. (2003). An examination of asynchronous communication experiences and perspectives of students in an online course: A case study. *Internet and Higher Education, 6,* 77-90.

Weissman, Julie, & Boning, Kenneth J. (2003). Five features of effective core courses [Electronic version]. *The Journal of General Education, 52(3),* 151-175. Retrieved January 28, 2006, from Wilson Web database.

Yena, Lauren, & Waggoner, Zach. (2003, Fall). One size fits all?: Student perspectives on face-to-face and online writing pedagogies. *Computers and Composition Online.* Retrieved December 21, 2005, from http://www.bgsu.edu/cconline/ yena-waggoner/index.html

Zhao, Chun-Mei, & Kuh, George D. (2004). Adding value: Learning communities and student engagement. *Research in Higher Education, 45,* 115-138.

9
SECURITY AND PRIVACY IN THE WIRELESS CLASSROOM

Mya Poe
Simson Garfinkel

[The] United States Attorney for the Eastern District of Virginia, announced that Myron Tereshchuk, 42, of Hyattsville, Maryland, has pled guilty to one count of attempting to extort $17 million over the Internet. . . . the defendant accessed the Yahoo account through unsecured wireless access points and the unauthorized use of the University of Maryland computer network and students' accounts. In the emails, the defendant demanded $17 million or he would disclose additional MicroPatent proprietary information and launch distributed denial-of-service attacks against intellectual property attorneys' computer systems worldwide.

(U.S. Department of Justice, 2004, ¶ 1 & 3)

The [University of Texas] administration issued a new policy . . . that bars students from running their own private Wi-Fi networks in campus housing. The unregulated hot spots are interfering with the university's own wireless service, which is offered freely to students and staff, campus technology administrators said.

(Borland, 2004, ¶ 1)

When wireless networks were introduced in 1990s, they promised to free users from the physical constraints of wired network settings. Indeed, wireless networks have freed writing instructors from dreary computer classrooms and have offered students exciting, new possibilities for digital composing. However, as writing programs move to wireless networks, Writing Program Administrators (WPAs) face new challenges in ensuring the security and privacy of program networks. As the cases at the beginning of this chapter illustrate, the illegal use of wireless networks not only raises knotty ethical and legal dilemmas about responsibility and negligence but also damages the trust among teachers, students, and the university. The purpose of this chapter is to help WPAs and composition instructors understand the scope of current security and privacy issues in wireless settings and foster a community of responsible wireless users. In this chapter, we discuss three kinds of wireless security issues that we think are most relevant to writing programs—controlling network access, handling interference from competing wireless networks and devices, and preventing harassment such as cyber-stalking.

Understanding the technology of wireless networking itself, however, is only partially the way to think about wireless security. The technological concerns of wireless security change rapidly, but the rhetorical and social implications of wireless security remain similar over time. Consequently, wireless security must also be thought of as a community value—one in which users should be able to participate in the development and implementation of security policies. By understanding security and privacy issues within this broader context, WPAs can work with faculty, students, campus Information Technology (IT), and the ombuds office to ensure that there is a clear, coherent policy on the ethical use of campus computing networks. More importantly, we can engage students in classroom (or digital) discussions about the role of individual responsibility in ensuring that wireless environments are safe spaces for productive exchange and learning.

COMPUTER SECURITY AND PRIVACY

Although computer security is a broad topic, ranging from password protecting laptop files to protecting national security, security goals are commonly articulated according to the Confidentiality, Integrity, and Availability model (National Institute of Standards and Technology Administration [NIST]). The Confidentiality, Integrity, and Availability model, which is shorted to the ironic acronym "CIA," has been under development by academics and practitioners in the computer security field since the 1970s. The CIA model frames three goals for security and privacy:

- Confidentiality requires "that private or confidential information *not be disclosed* to unauthorized individuals."
- Integrity requires that data *not be altered* without proper authorization.
- Availability means that systems should be *ready to use when wanted or needed.* (NIST, p. 7)

WPAs can use the CIA goals as guidelines for writing program wireless networks. By meeting those goals, WPAs can also help ensure that the program is meeting existing Family Educational Rights and Privacy Act (FERPA) regulations. For example, any student who uses a wireless network on a program network should have confidence that his/her written work will not be disclosed to unauthorized parties, that his/her work will not be altered without consent, and that s/he can access the network as needed to complete writing assignments.

In meeting the CIA goals, it is best to start with policy questions. Who should have access to the network: only authorized users of the writing program or anyone at the university with a wireless device? Should the wireless network be protected or should each computer on the network be responsible for protecting itself? Do the proposed policies allow for pedagogical innovation while preserving privacy and security? How will these policies be articulated to faculty and students, and how will such policies be enforced? The answers to these questions can help writing programs select the security measures that are best suited to the program's needs and goals.

OPEN FOR ACCESS

In considering who should have access to a program's wireless network, it is tempting to want completely open access. Indeed, open access is great for usability: If your laptop, Personal Data Assistant (PDA), or other wireless device can receive a wireless signal, you can access the Internet. Of course with open access, hackers can also gain access. In 2003, for example, a hacker broke into the University of Kansas wireless system and downloaded personal information on 1,450 international students—information that the universities had been forced to collect as part of a federal counterterrorism project (*USA Today*, 2003). George Mason University suffered a computer breach in 2005 when a hacker accessed personal information including Social Security Numbers for 32,000 people (*The Washington Post*, 2005). Malicious hackers who access a poorly secured network can wreak havoc with the information they "release." The University of Maryland found this out the hard way in early 2004 when Myron Tereshchuk gained unauthorized access to student accounts and then used the accounts to send harassing e-mails to the customers of MicroPatent, LLC, in an attempt to extort $17 million from the company (Poulsen, 2004).

There are four main technological "solutions" to control access to the program's wireless network so that incidents like these don't occur: restrictions based on MAC address registration; the use of wireless encryption protocols such as WEP and WPA; the use of Virtual Private Network protocols; and the use of end-to-end encryption. We outline the benefits and drawbacks of these four methods to better allow a WPA to make an informed decision about, or to argue for, the best form of network security for his or her program. Although computer security and privacy technology changes rapidly, these four methods are representative of ongoing discussions within the field of computer security and privacy about who, where, and when to implement security measures to maintain the CIA goals for safe computing.

MAC Address Registry

The simplest but unfortunately the least effective technique for securing a wireless network is to have students register the MAC address of their wireless device. A *MAC address* is the unique 48-bit serial number that a manufacturer assigns to each wireless device it produces. MAC addresses are commonly displayed as a set of 12 hexadecimal digits grouped and separated by colons, for example, 00:0a:95:f2:40:42. When students register their device's MAC address, the network only provides access to devices whose MAC addresses are on a predetermined list.

Restricting access to a network by MAC address is an effective tool for preventing casual access to the network, but it is not particularly effective against hackers. There are two reasons for the ineffectiveness of MAC address restriction. First, MAC address restrictions do not prevent devices from *receiving* data that are sent over the network. As a result, even with MAC address registration, a hacker is able to receive all of the data that is sent over the network; he or she just can't send any data over the wireless network. The second reason that MAC address restrictions are ineffective is that MAC addresses can be changed through simple revisions of a computer's configuration.

Although the registration of MAC addresses is problematic, many universities use it as their choice of wireless security. From Casper College in Wyoming (2004) to the University of Melbourne (2005), students are required to register the MAC address of their wired and wireless devices. For such universities, MAC address registration provides a useful compromise between usability and security—it provides some security with minimal hassle to users. Because most users do not know how to change the MAC address of their computers, only allowing devices with a registered MAC address to use a wireless network prevents access by most unauthorized users.

Cryptographic Protocols

An improvement over MAC address registration is the use of cryptographic protocols that scramble information as it is sent through the air. Two common cryptographic protocols are WEP (Wireline Equivalent Privacy) and WPA (WiFi Protect Access). The main use of these systems is for access control: if hackers do not have the necessary encryption key (password), they will not be able to use the network.

Password-protected wireless networks, such as provided by WEP and WPA, typically don't work for writing programs because students may distribute the password to the program's wireless network to their friends. One way that universities have tried to limit such sharing is through policies. For example, the computer usage guidelines at Carnegie Mellon University states that students are not permitted to share encryption keys:

> Sharing your password or account with the specific exception of staff or faculty members allowing their support personnel to access their accounts in order to provide services appropriate to their job functions. Note that some policies for the accessing of specific systems or data . . . explicitly forbid the sharing of passwords used to access them, and that such restrictions for those specific systems override this policy. (Carnegie Mellon University, 2003, Misuse and Inappropriate Behavior section, ¶ 4)

Although such policies are straightforward, students are unlikely to follow them. As a result, policies about password sharing need to be combined with other measures to ensure that the wireless network is secure. A far more significant problem with relying on WEP and WPA for securing wireless networks is that they do not provide for wireless privacy between different users on the same wireless network. In writing programs where teachers may share electronically graded projects or other private materials, such access may lead to an unintended distribution of student work, confidential e-mails, or student grades. In other words, once a user has access to the wireless network, he or she can eavesdrop on the communications of other users because there is no "isolation" between users.

Virtual Private Networks (VPNs)

One way to provide secure sharing between wireless users is to use Virtual Private Network (VPN) technology to give each wireless user an encrypted "tunnel" for all of his or her communications. A VPN tunnel connects from a user's laptop to a secure VPN server that is located somewhere on the university's wired network. Because each user has his or her own tunnel, other users cannot decipher the contents of wireless communications. One of the big advantages of using VPNs is that the cryptographic tunnel can protect users anywhere in the world—a professor or student traveling in Pakistan, for instance, can open up a "tunnel" to the campus network from a cyber café and use campus services as if he or she is sitting in the office.

Universities such as Iowa State (2004), Northern Arizona University (2003), and Denison University (2005) encourage the use of VPNs. The Denison University policy states:

> Wireless is a "shared resource." This means that unless your connection is encrypted, it is entirely possible that someone could be reading your data. To ensure that your connection is secure and encrypted, we encourage you to use a Virtual Private Network (VPN) client. If you do not use VPN, be aware of the dangers of using a shared resource like wireless, and avoid sending personal or sensitive data across the wireless network. (Denison University, 2005, How Secure is Wireless Networking section, ¶ 1)

Universities such as Denison offer software on their campus computing web pages so that students and faculty can create their own VPNs. On such campuses, it is just assumed that the university wireless network is insecure, and rather than trying to ensure that the wireless network is secure, the burden of security is placed on the user. The benefit of placing security on users is that it releases writing programs from assuming the entire responsibility for wireless security and privacy. The drawback, however, is that this approach requires users to configure the VPN and possibly to install additional software.

In the end, although Virtual Private Networks offer more security and privacy than found on open wireless networks, they still are not ideal for confidential information because they only encrypt the data while data are in transit. Once the data are received by the endpoint, they are stored without encryption (Garfinkel, 2003). Likewise, information that is sent over the Internet to an outside source will not be protected with a VPN. As a result, WPAs may consider VPNs useful for ensuring that students can get secure access to information on the university server but not secure for sending communications sent outside the university.

End-to-End Security

The best way to ensure security over a wireless network is through end-to-end security. Simply put, end-to-end security means that information is protected from the server, through the Internet, to the user—from one end of the Internet connection to the other. One example of end-to-end security is downloading a list of student registrations from a university website that uses the Secure Socket Layer (SSL) encryption protocol, as evidenced by URLs that begin "https:" instead of "http:". Many security experts believe that end-to-end security provides the highest level of security because it puts security under control of the endpoints—the computers that are actually depending upon the security to function properly (Saltzer, Reed, & Clark, 1984).

Although SSL is commonly used today for web pages that send credit card numbers, it is not commonly used for securing other web pages or for protecting most passwords, such as the passwords used to download e-mail. Some campuses, however, are beginning to encourage users to use SSL for webmail or e-mail. For example, the University of Arizona (2005) encourages users to use SSL connections for webmail and IMAP (a program like SSL) for e-mail. We believe that SSL or a similar technology is a suitable security layer for all e-mail and courseware web sites because it ensures security for users over the entire wireless connection.

ROGUE WIRELESS

Although accessing unsecured networks and stealing university information is the most commonly reported form of security and privacy breach in
wireless settings, there is another more benign but equally as problematic
problem for campuses—*rogue networks*. Rogue networks are wireless networks that are not part of the official university network. Because wireless
networks run on a finite amount of bandwidth, rogue networks and other
wireless devices such as Bluetooth technology "compete" with that limited
bandwidth and can jam the main wireless network.

Rogue wireless networks are common on college campuses. Consider
that while driving around the University of New Hampshire, one researcher
found close to 50 such wireless networks in less than two hours (Murphy,
2003). In order to maintain control over the wireless channels, some universities such as George Washington University have banned rogue network devices:

> The establishment of wireless networks is not authorized in the resi-
> dence halls due to security concerns. . . . Establishing a wireless net-
> work or using a wireless access point within a residence hall may be
> subject to judicial action and permanent loss of your ResNet connec-
> tion. (George Washington University, n.d., Wireless in Residence Halls
> section, ¶ 1)

Other universities have taken a more moderate approach to rogue networks and competing wireless devices. For example, Rutgers University
just asks that students register them:

> In the residence halls, students may install wireless LAN systems with-
> out any special permission. They must be registered in the registration
> database. Students are expected to work with each other to deal with
> interference. While OIT staff will not in general manage channel allo-
> cation in the residence halls, they may intervene if a particular wireless
> installation is being operated in a manner that unreasonably interferes
> with other users, if an installation interferes with University-operated
> installations, or in support of student-led initiatives to coordinate allo-
> cation of channels. (Rutgers University, 2005, Policy Statements,
> Residence Halls section, ¶ 1)

We do not recommend that writing programs set up their own rogue networks apart from the official university wireless network. Setting up unofficial networks is not only a violation of many campus computing policies, but it's also interpreted by some as being a "bad" wireless citizen. Instead, writing programs can work with campus the IT department and other departments to ensure that the wireless needs of the program's students and faculty are being met. Coordinated planning, in fact, can be made part of the university-wide computing mission. For example, Rutgers University recommends the following:

> In order to allow all units to have access to wireless LAN technology, it may be necessary for some units to adjust their behavior to make more efficient use of channels. For example, if one unit has a large number of access points in individual offices, these might exhaust the available channels. It would be reasonable to ask such a unit to replace these individual access points with a more coordinated approach. It may often be advantageous for all the units in a building to do a single build-ing-wide wireless system. (Rutgers University, 2005, Policy Statements, ¶ 7)

We also recommend that programs check for rogue networks in the vicinity of writing classrooms, especially if the university does not require students to register MAC addresses. A rogue network that is not active at the start of a class can become very active as the class progresses, causing progressively more interference. The common way of contending with rogue access points is to scan for them and disconnect them once their origin has been located. Many universities, such as the University of Massachusetts and Syracuse, make it clear in their policies that they have the right to disconnect any competing wireless devices or rogue networks from the main network:

> The University reserves the right to remove or disconnect any access point not installed and configured by OIT personnel or specifically covered by prior written agreement and/or arrangement with OIT. (University of Massachusetts, 2003, Wireless Airspace Policy, ¶ 8)

> If network disruptions occur the University CIO reserves the right to disable any access point or wireless device not installed, configured, or approved by CMS personnel. (Syracuse University, n.d., Wireless User Standards, ¶ 6)

As the price of a wireless technology becomes cheaper, rogue networks and interference from other wireless devices will become more common on university campuses. Writing programs should plan for substantial demands on their wireless network and prioritize those demands if they are to ensure that faculty and students continue to have good access.

WIRELESS HARASSMENT

Although the potential for harassment is present in any digital writing environment, the security vulnerabilities of today's wireless network technology make harassment easier to commit and more difficult to trace. Some of the most common forms of wireless harassment are covert monitoring, denial of service, and overt tracking. To help a WPA, instructors, and students deal with these problems, we identify each form and suggest possible strategies for contending with these instances.

Covert Monitoring

Because most wireless networks do not employ end-to-end encryption, it is usually possible for one user to "eavesdrop" on the communications of another user. Monitoring tools, which are freely downloadable on the Internet and can run on ordinary laptop computers, typically intercept every "packet" that is sent over the wireless network and copy the information on to the intruder's computer. The intruder can then reassemble the information to see what the victim was viewing online. For example, at some network security conferences, computers in the hallway eavesdrop on the wireless network and display randomly chosen images from the network traffic as their screen savers. People passing in the corridor can look at these computers and get a sense of what their fellow conference attendees are browsing on the Internet.

Although women historically were the most likely targets of such harassment (Spertus, 1991), these days international students also have reason to be concerned about covert monitoring. For example, a list of websites visited by a foreign student might be given to authorities with the notation that the student was visiting websites belonging to militant organizations. Although possibly true, it might be the case that all of the students in the class had been instructed to read the web site by their teacher as part of a study on political rhetoric or as research for a writing project.

Although some universities such as MIT have network policies that specifically prohibit students and faculty to monitor network traffic, such policies cannot stop monitoring; they simply provide a "rule" that can be used against offenders if they are caught. One warning sign of possible monitoring is when one student seems to have uncanny knowledge about the actions or interests of another student. Such a sign should alert faculty members to the possibility of covert monitoring. Although we assume that most writing students and faculty are not participating in this form of harassment, we know that WPAs must have information about the technological capabilities of the resources they provide their faculty and students.

Disruption

Once a student monitors his or her fellow classmates, that student may be tempted to go further and actually disrupt another student's usage. Such actions are sometimes called *denial of service* because the victim is denied access to the network by the intruder. The most simple and straightforward approach to disrupt service is for an intruder to send a single packet to the victim's computer that makes the victim's computer lose its connection to the network. An intruder can also inject false web queries over the wireless network so that a student who attempts to go to a course-related website arrives at a pornographic website instead. It is even possible to modify the content coming from remote websites to the computers of a single user or of the entire class. This form of harassment can effectively undermine both the sense of well being in a classroom and undermine the learning community.

Denial of service attacks can be detected using network analyzers. There is, however, another "low tech" way to identify potential denial of service attacks. If a particular student repeatedly has unexplained problems with his/her computer in the writing classroom but nowhere else on campus, it may be useful to consider that a denial of service attack is occurring.

MAC Address Tracking

Yet another approach to wireless harassment is to track a user's wireless device as he/she moves throughout the university campus. Because wireless devices constantly transmit their MAC address whenever the wireless card is in use, a user's location can be monitored through his/her laptop's MAC address. A network of receivers scattered throughout the campus can silently collect this information, record it, and build a profile of when individual users come and go.

Newbury Networks, a Massachusetts-based company, has developed a product that uses this capability to create a system for tracking users of handheld computers as they walk around museums and businesses. The system triangulates wireless users using their MAC address and their wireless signal. Museums can use it to display different web pages or maps on a handheld computer as a person moves from exhibit to exhibit. The system is accurate to within three meters (Garfinkel, 2002). In November 2005, MIT demonstrated a campus-wide wireless tracking system in the MIT Museum (Brooke, 2005). Maps of the Institute were shown on a large screen with red dots indicating the presence of individuals with wireless devices. Although the dots did not show the student names, this information could have been readily displayed by simply consulting the MAC address registry.

MAKING WIRELESS MORE SECURE AND MORE EQUITABLE

As demonstrated through the various campus computing policies, universities are taking a variety of approaches to ensure that wireless security is guarded through both technology solutions and policy. At the end of the day, wireless computing security means that there must be a larger university-wide commitment to the issue, in the same way that universities as a whole have policies about plagiarism and copyright infringement. Although the security of the wireless network at large is the responsibility of computer services, writing programs can help ensure that the university is living up to its responsibilities. Our experience indicates that many universities are not, despite the fact that their networks are being used to transmit information that is entitled to the highest level of protection under both Federal law and the information policies of most universities. It is disturbing that students are being asked to use wireless networks without being given the assurance that their legal rights are being met.

On a positive note, a review of university computing policies shows that within the past several years, most major universities have established specific wireless computing policies. Computing codes of ethics, such as those at Carnegie Mellon University, are part of the university's policy statements positioned along with their policies on plagiarism, sexual harassment, intellectual property, free speech, and copyright. Violations of wireless policy involve the same kinds of redress used in other kinds of ethics violations. For example, the Carnegie Mellon University policy states:

> Inappropriate behavior in the use of computers is punishable under the general university policies and regulations regarding faculty, students and staff. The offenses mentioned in this policy range from relatively minor to extremely serious, though even a minor offense may be treated severely if it is repeated or malicious. Certain offenses may also be subject to prosecution under federal, state or local laws. (Carnegie Mellon University, 2003, Enforcement section, ¶ 1)

However, putting policies in place will not work unless faculty and students read them and understand their personal responsibility to wireless security. One approach is for a WPA to consult with staff and faculty to develop reasonable and rhetorically sound policies related to wireless use. This consultation, in the form of open meetings, focus groups, and other community sessions, can help to inform faculty about wireless technologies and enable them to create innovative, secure wireless pedagogies. Teaching assistants who teach in wireless classrooms must also be taught about security and privacy rights and responsibilities. Another approach is to assign students the task of researching, reading, and perhaps even writing wireless policies. Although many students may have used wireless technology in their high schools, it was unlikely that they were involved in deciding any of the IT policies regarding wireless computing. For example, what is the purpose of wireless policy? How does policy affect students and their work in the course? Is the recourse provided to the University under the policy clear, appropriate, and purposeful? How does this policy fit within the larger rules, responsibilities, and relationships of the writing classroom and the university community?

Engaging students in the rhetorical aspects of policy will also help establish a dialogue with students about policies, their potentials and limitations, and their role in a democratic society. For example, what does it mean that many universities ban covert monitoring by students and faculty while themselves reserving the right to monitor traffic without consent of users? Why do universities get to decide the appropriate measures of wireless security and privacy without involving users in those decisions? What does it mean that those decisions are usually left to a small group of "experts"? Why is wireless security so often haphazard when wireless computing is mandatory in many schools? And what are the implications when the highest levels of security are only found at the wealthiest universities? Critically examining a university policy such as the following one from Denison University may help students begin to think about power relations in technological settings:

> The university may specifically monitor the activity and accounts of individual users of university computing and network resources, including individual login sessions, content and communications, without notice. . . . The university, in its discretion, may disclose the results of any such general or individual monitoring, including the contents and records of individual communications, to appropriate university personnel or law enforcement agencies under the direction of a court of law and may use those results in appropriate university disciplinary proceedings. (Denison University, 2005, Security and Privacy section, ¶ 1)

Having students work with the language of the policy itself can provide a real rhetorical situation for their own writing. Many campus computing policies are written in convoluted, legalese prose. For example, compare the following two policies:

> Unless steps are taken to protect them, wireless LAN installations are open to anyone within range of the access point. If a wireless access point is connected to the Rutgers network without restrictions, anyone with the proper equipment will be able to access the Rutgers network, even from outside the building. Furthermore, anyone with the proper equipment can spy on traffic. They can see users' passwords as well as other data. As Rutgers moves more and more services online, the amount of damage that can be done by unauthorized people learning passwords of Rutgers users is increasing. . . . These dangers are not just theoretical: Tools to tap nearby wireless networks are widely available, even for palmtop devices. A whole subculture has sprung up of people going around, scanning for open wireless nodes, and publicizing them to people who want free wireless access. Interference among installations is already visible in several buildings at Rutgers. (Rutgers University, 2005, ¶ 2)

> The University of Arkansas wireless networking systems is an enterprise system covering external areas of the campus and interconnected building wireless networks. These systems are intended to allow campus users access to all campus computing facilities from mobile or portable computers such as notebook computers and personal digital assistants (PDAs). A portion of radio airspace on the campus serves as the trans-

port medium for a part of the campus network. Accordingly, certain computing, networking, and security policies apply to that segment of the radio airspace on campus. (University of Arkansas, 2004, ¶ 1)

What are some of the effects of the language used in wireless policies on readers? How does it change notions about the ownership of wireless security? How do the policies position faculty, students, and staff? What assumptions are reflected in the language of policy? Having students write security policies in their own language or interview their friends about wireless security issues and then craft their own policies helps students see the relationship between language use and power. Composing their own policies also helps students take ownership of wireless security.

Such teaching opportunities engage students in the lived experience of wireless computing, making wireless security less a matter of managing technological innovation and more about the ways in which communities use and value notions of privacy and security in technological spaces. Such student education combined with wise choices by WPAs and faculty will help ensure that writing programs participate in both technologically secure network access and communities of ethical wireless computing.

REFERENCES

Borland, John. (2004, September 9). Students, college face off over Wi-Fi. *ZDNet*. Retrieved December 15, 2004, from http://news.zdnet.com/2100-3513_22-5360510.html

Brooke, Donald. (2005, November 3). MIT maps wireless users across campus: System allows students to see where everyone is at any time. Retrieved November 10, 2005, from http://www.msnbc.msn.com/id/9914807/

Carnegie Mellon University (2003, May 16). Carnegie Mellon university computing policy. Retrieved November 1, 2005, from http://www.cmu.edu/policies/documents/Computing.htm

Casper College. (2004, August 16). Register MAC address. Retrieved November 1, 2005, from http://acadcomp.caspercollege.edu/wireless/wirelessReg.aspx

Denison University. (2005, November 15). FAQs about wireless. Retrieved December 31, 2005, from http://www.denison.edu/computing/wireless/#secure

Denison University Information Resources Advisory Committee. (2005, April 27). Security and Privacy. Retrieved December 31, 2005, from http://www.denison.edu/irab/aup/

FBI investigating theft of data on international students by hacker. (2003, January 24). *USA Today*. Retrieved December 31, 2004, from http://www.usatoday.com/tech/news/2003-01-24-records-hacked_x.htm

Garfinkel, Simson. (2002, May 27). Wi-Fi 'hot spots' allow laptop, PDF users to be covertly tracked. *The Seattle Times*. Retrieved January 29, 2005, from http://www.simson.net/clips/2002/2002.SeaTimes.05-28.WiFiTracking.htm

Garfinkel, Simson. (2003, January). On the same wavelength. *CSO Magazine*. Retrieved January 28, 2005, from http://www.csoonline.com/read/010903/machine.html

George Washington University. (n.d.). ResNet code of conduct. Retrieved January 15, 2006, from http://gwired.gwu.edu/resnet/internet/resnetcode/

George Mason officials investigate hacking incident. (2005, January 13). *The Washington Post*. Retrieved January 17, 2005, from http://www.washington-post.com/wp-dyn/articles/A5188-2005Jan12.html

Iowa State University. (2004, July). Using a virtual private network (VPN) with the Iowa state network (windows 98, 2000, xp). Retrieved November 10, 2005, from http://www.it.iastate.edu/pub/wng334/wng_334.html

Murphy, Sean. (2003, November 16). Wireless computers are the next wave in technology, but a new batch of hackers may be right down the road. *Foster's Sunday Citizen*. Retrieved December 15, 2004, from http://premium1.fosters.com/2003/news/nov_03/november_16/news/bu_1116b.asp

National Institute of Standards and Technology Administration, U.S. Department of Commerce. (n.d.). *An Introduction to Computer Security* (Publication No.800-12). Retrieved November 1, 2005, from http://csrc.nist.gov/publications/nist-pubs/800-12/handbook.pdf

Northern Arizona University. (2003, January 31). NAU's virtual private network (vpn) service. Retrieved November 1, 2005, from http://www4.nau.edu/its/vpn/Noguchi, Yuki.

Poulsen, Kevin. (2004, June 25). Wi-fi hopper guilty of cyber-extortion. *Security Focus*. Retrieved December 31, 2004, from http://www.securityfocus.com/news/8991

Rutgers, The State University of New Jersey. (2005, May 23). Wireless lan policy. Retrieved November 1, 2005, from http://oit.rutgers.edu/wireless-policy.html

Saltzer, Jerry H., Reed, David P., & Clark, David D. (1984, November). End-to-end arguments in system design. *ACM Transactions on Computer System* (pp. 2, 4, 277-288). Retrieved November 1, 2005, from http://doi.acm.org/10.1145/357401.357402

Spertus, Ellen. (1991). Why are there so few female computer scientists? *MIT Artificial Intelligence Laboratory Technical Report #1315*. Retrieved January 29, 2005, from ftp://publications.ai.mit.edu/ai-publications/pdf/AITR-1315.pdf

Syracuse University. (n.d.). Wireless user standards. Retrieved December 1, 2005, from http://cms.syr.edu/connecting/wireless/userstandards.cfm

University of Arizona. (2005, November 3). Laptop/wireless security. Retrieved November 7, 2005, from http://www.cs.arizona.edu/computer.help/policy/Laptop_Wireless_Security.html

University of Arkansas. (2004, September 7). University of Arkansas wireless airspace policy. Retrieved November 1, 2005, from http://compserv.uark.edu/policies/airspace.htm

University of Massachusetts. (2003, August 15). Wireless airspace policy. Retrieved November 1, 2005, from http://www.oit.umass.edu/policies/wireless.html

University of Melbourne. (2003, February 3). MAC address registration. Retrieved May 5, 2005, from http://www.cs.mu.oz.au/register/

U.S. Department of Justice. (2004, June 8). News release. Retrieved December 31, 2004, from http://www.usdoj.gov/criminal/cybercrime/tereshchukPlea.htm

PART IV

Teaching and Learning in Motion: Mobility and Pedagogies of Space

10

PERPETUAL CONTACT

RE-ARTICULATING THE ANYWHERE,
ANYTIME PEDAGOGICAL MODEL OF MOBILE
AND WIRELESS COMPOSING

Amy C. Kimme Hea

In *Technology and Literacy in the Twenty-First Century: The Importance of Paying Attention*, Cynthia L. Selfe (1999c) carefully traces the roles of government, education, business and industry, parents, and ideology in forging the connections between literacy and technology. Much like the other critical precursors to this extensive project (Hawisher & Selfe, 1991; Kaplan, 1991; Selfe, 1999b), Selfe's book argues that simplistic, deterministic assumptions of technologies—as either our boon or bane—mask our ability to think carefully and ethically about the integration of technology, its impact on literate practices, and our ability as teachers and scholars to work toward meaningful change. Wireless and mobile technologies, although not new, are being touted as the next great champion against all educational woes. Such deterministic claims for positive, uncritical educational reform through wireless connections are echoed in the U.S. Department of Education's special 2004 National Education Technology Plan report entitled, "Toward a New Golden Age in American Education—How the Internet, the Law and Today's Students Are Revolutionizing Expectations." Introduced as one of the emerging models for better education, laptop initiatives such as the one in Henrico County Public Schools in Virginia are promoted as a best technology practice to increase student performance (U.S. Department of Education, 2004 ¶ 19). Quoted on Apple's educational success website, Dr. Mark Edwards, Superintendent of Henrico County Public Schools, stresses that

> [w]ith our iBook laptop program, instead of having one hour a week in a computer lab, our students now have the opportunity for continuous learning, 24/7. They're using their iBook computers on the school steps, on the playground, at home . . . everywhere they go. The wireless world, along with the portability of the laptops, and the dynamic nature of the digital content are creating a synergy that will have a huge impact on schools and classrooms. (Apple Computer, Inc., 2003, ¶ 1)

As new educational reforms urge teachers and administrators to move toward wireless and mobile configurations, we are reminded of the pressing need to be critical of both the articulations of technology and literacy and the construction of student and teacher subjectivities being advocated. Dr. Edward's continuous learning model relies on the concept of ubiquity— or the existence, or apparent existence, of technology being everywhere, pervasive and immersive. Constructed discursively and practically in relation to technology, ubiquity is necessary if we are to believe that access to a constant stream of content increases student literacy. A faith in ubiquity also provides for the unwieldy construction of the 24/7 student and the on-demand teacher. The "huge impact" Dr. Edward's alludes to has yet to be considered carefully as the pace to integrate wireless and mobile pedagogies increases.[1]

Assuming the position that wireless laptops and mobile learning (m-learning) are clearly beneficial, always desirable, and capable of bridging the digital divide leaves us little room to develop critical wireless and mobile pedagogies. In effect, such determinisms serve to reify existing social inequities and inscribe rigid subject positions for students and teachers. The double articulation of ubiquity and mobility attempts to construct all spaces—local and global—in relation to the goal of complete, uninterrupted connectivity. Rather than vilify measures to increase connectivity, I call for an analysis of the pedagogical implications of anywhere, anytime educational practices. To question the deterministic relationships between ubiquity and mobility, I first examine ubiquitous computing, specifically the developments and discourses of the calm technology and Oxygen projects. Such ubiquitous computing projects are intimately connected to the assumption that wireless and mobile technologies will foster a natural learning environment, erase boundaries of learning for students and teachers, and free all of us from the failings of our current educational systems. These determinisms are intertwined with spatial relations, particularly non-place theories as defined by Paul Virilio's *dromology*—the study of speed—and Marc Augé's views on the overabundance of spaces. After reviewing these spatial theorists, I critique the constructions of literacy and teacher and student subjectivities as defined through two wireless laptop initiatives: my own local school district's 1:1 laptop program at Empire High in Tucson, Arizona, and Massachusetts Institute of Technology's (MIT) global program of One Laptop per Child (OLPC). Both these local and global programs uncritically embrace the technological determinisms of ubiquity and mobility. To counter a space-less perspective devoid of rhetorical agency, I offer some possible starting points for writing program administrators (WPAs) and composition teachers to develop informed wireless and mobile practices for teaching and learning. Without a rhetorical perspective on wireless and mobile initiatives, we run the risk of assuming literacy is natural and neutral, technology has agency, and student and teacher access to technology erases all political, social, and cultural relations.

CULTURAL NARRATIVES OF UBIQUITY

A critical examination of wireless and mobile technologies must take into account the ways these technologies construct and are constructed by dominant educational discourses. Notions of 24/7, just-in-time, on-demand, and m-learning assume that wireless and mobile technologies are ubiquitous. Ubiquitous computing (ubicom) was first defined as "thousands of highly distributed, interconnected, often invisible computers, blended into the natural environment, operating without engaging peoples' conscious senses or attention" (Weiser, 1991, p. 94). Ubicom may best be demonstrated through Xerox Parc's calm technology research and MIT's Oxygen project. These projects provide a means to understand ubiquity and its connection to deterministic wireless and mobile initiatives. Ubiquitous computing, in fact, argues for the invisibility of technology—making critique of technological practice nearly impossible or irrelevant—and perpetuates the idea that individuals need not consciously engage technology—assuming that agency in relation to technology is unnecessary and undesirable.

Ubiquitous computing research is often credited to research technologists working in the 1990s at Xerox Parc Palo Alto Research Center. Mark

Weiser and John Seely Brown, wrote a now-famous essay, "The Coming Age of Calm Technology," on the prospects of ubicom. In this piece, Weiser and Seely Brown (1996) argue that ubiquitous computing will "require a new approach to fitting technology in our lives," an approach they term "calm technology" (¶ 2). Because ubiquitous technologies may lead to cognitive and sensory overload if ill-designed, Weiser and Seely Brown (1996) emphasize three principles of calm technology: (a) motion between center and periphery, (b) peripheral reach, and (c) locatedness (¶ 19-26). The first two principles have to do with our ability to give, maintain, and shift our attention and concern. Being able to categorize our experiences as either central or peripheral and then push certain cognitive data to the periphery, allowing for a cognitive and sensory respite, will be necessary for our immersion in ubiquitous technologies (Weiser & Seely Brown, 1996, ¶ 22). Locatedness is concerned with "details" that make us feel connected (Weiser & Seely Brown, 1996, ¶ 23). These principles have to do with sifting through information and positioning ourselves within the openness of ubicom technologies. Within this framework, all relations are spatial—a point I will return to in my upcoming discussion on nonplace theory. Weiser and Seely Brown (1996) go so far as to suggest that "[t]he 'UC' [ubiquitous computing] era will have lots of computers sharing each of us" (¶ 9).

Another illustrative vision of ubiquitous computing is MIT's Computer Science and Artificial Intelligence Laboratory's Oxygen project, which was featured in the August 1999 edition of the popular science magazine, *Scientific American.* The Oxygen project seeks to make technology as ubiquitous as the element that we need to live—oxygen. One of the lead scientists, Michael Dertouzos claims, it is about "bringing abundant computation and communication, as pervasive and free as air, naturally into people's lives" (Manjoo, 2001, ¶ 15). Through a virtual network of personal computer devices, the Oxygen project hopes to integrate technologies fully into our lives and tailor them so that they learn our habits, routines, and preferences to such a degree that we are being engaged by them rather than us engaging them. As Dertouzos (1999) asserts, "people who want to exploit the new world of information should explore the capabilities of the new Oxygen technologies" (p. 53). He further stresses that "[e]very individual and organization will have access to these resources and that the ones who will truly do more by doing less will be the ones who learn how to integrate these technologies and their people into a well-oiled, humming whole" (Dertouzos, 1999, p. 53). Prophetically, the Oxygen group wishes to make all of us a part of a universal network where our experiences are transparently and seamlessly negotiated through a shared computation and communication system.

Although the calm technology and Oxygen projects may sound oddly like science fiction, the next generation Internet protocol, Internet Protocol Version 6 or IPv6—which was tested at the time of Weiser and Seely Brown's article and has been under implantation since 2004—can address more than 1,000 devices for every atom on the earth's surface. Embedding microprocessors and coupling them with Internet capabilities means that your toaster, blender, radio, refrigerator, car, walls, clothing, and any other objects can begin communicating with one another. In 1996, Weiser and Seely Brown explained that the capability of such communication devices could allow for "[c]locks that find out the correct time after a power failure, microwave ovens that download new recipes, kids toys that are ever refreshed with new software and vocabularies, paint that cleans off dust and notifies you of intruders, walls that selectively dampen sounds" (¶ 12). All of these functions, however, remain relatively invisible to individuals— on the periphery—to use Weiser and Seely Brown's own term. These two projects reveal a similar—and disturbing— assumption that our best technologies have us, know us, and engage us without us being consciously aware of them.

Without slipping into technological determinism myself and arguing that these visions will necessarily become reality,[2] I want to suggest a much more important point: Regardless if any of these projects come into existence, current discourses and practices of already available wireless and mobile technologies—such as Wi-Fi, Personal Data Assistants (PDAs), and wireless laptops—are constructing expectations for how students and teachers should learn and teach. Within such frameworks, our relationships with students will be mediated through invisible networks, information will be translated to serve their needs and our own, and decisions will be made without much conscious effort.

Currently, agency, critical literacy, and student and teacher subjectivities are either being erased or defined in narrow, circumscribed ways through ubicom constructions. Such constructions are exemplified in IBM's ThinkPad University Program. In its promotional video, "Education on Demand," IBM offers student and teacher testimonials about their access to one another and their wireless IBM ThinkPad laptops. In one such testimonial, a student stresses that "teachers can get in touch with you any time of the day," and still another student explains that "I needed help while I was doing my laundry, and so I emailed my teacher, and she got back to me right away while I was still in the laundry" (IBM, 2003). One professor on the video argues for the benefits of the program explaining that "I talk with students 24/7" (IBM, 2003). Such assumptions about perpetual student-teacher contact tend to emphasize management issues rather than complex learning situations. IBM's promotional video also features a student who lauds his university stressing that "a lot of people can't afford computers on a regular basis you know. If it wasn't for Morrisville [State College], a lot of us wouldn't have a laptop. They did us a favor with that" (IBM, 2003). It is significant that this grateful student is a person of color making the "us" an even more problematic term. In a similar move to erase difference, Gateway Computers (2004) emphasizes "Your Students Come from Diverse Backgrounds, Which Makes Universal Access More Important Than Ever. Gateway Makes One-To-One Computing Easier Or More Affordable Than Ever Before" (p. 15). These IBM and Gateway promotions use the discourse of ubiquitous computing to argue that their wireless laptops can serve our needs and transcend boundaries, even those of time, space, and diversity.

SPATIAL THEORIES: ACCELERATION AND OVERABUNDANCE

As noted, many wireless programs embrace the idea that space and time can be transcended or even collapsed through the integration of laptops and other portable technologies such as PDAs, iPods, and cell phones. In effect, an uncritical perspective on space-less and time-less technologies leads to an erasure of history, place, context, and agency. The ahistorical, placeless, timeless, agentless discourse is replete with dangers: Literacy and subjectivity are seemingly neutral, as are the technologies that contribute to the production of both. This neutrality privileges dominant social classes, making issues of resistance moot. Context seemingly melts away, and the role of rhetoricians and writers is to speak a single, universal truth. Rhetoric and composition scholars have discussed the importance of spatial relations as part of our research practices (Kimme Hea, 2008; Porter & Sullivan, 1997), institutions (Porter, Blythe, Grabill, & Miles, 2003; Porter, Sullivan, Blythe, Grabill, & Miles, 2000), composition theories (Mauk, 2003; Reynolds, 1998, 2004), rhetorical history (Mountford, 2001), and electronic spaces (Johnson-Eilola, 1997, 2005; Payne, 2005; Selfe & Selfe, 1994). Many of these scholars have adeptly articulated spatial relations drawing upon Henri Lefebvre (1991), Michel DeCerteau (1984), Michel Foucault (1978, 1980, 1984, 1997), and perhaps most often Edward Soja (1989, 1996). Whereas these critical perspectives on spatial relations discuss the making of and being made through space, I want to extend the conversation to the placelessness of wireless and mobile technologies. Through the seeming eradication of space (and time), wireless and mobile discourses construct themselves as omnipresent, invisible and everywhere.

One dimension of placelessness is the phenomenon of time and space collapsing because of technological acceleration. This collapse is a returning motif in urban planner and critical theorist Paul Virilio's scholarship. In fact, Virilio (1986) claims that "the reduction of distances has become a strategic reality bearing incalculable economic and political consequences, since it corresponds to the negation of space" (p. 133). Virilio effectively argues for the serious study of speed, which he calls *dromology*, and the ways that channels of acceleration (from roads to the Internet) can impede critical inquiry. On this point, we rhetoricians and compositionists may want to take note: Wireless and mobile technologies are constructed around the idea that access to information is quick and easy. Like the rhetoric of ubicom, wireless technology is argued to be always, already around us, accommodating our needs, and teachers and students are positioned as only benefiting from instant access to perpetual, seemingly limitless information.

Taking a slightly different stance from Virilio's collapse of time and space, Marc Augé focuses more attention on the channels between spaces. Augé argues that we have not lost space but rather we have an overabundance of spaces, many of which he calls non-places. For Augé these non-places "designate two complementary but distinct realities: spaces formed in relation to certain ends (transport, transit, commerce, leisure), and the relations that individuals have with these spaces" (1995, p. 94). Whether by accident or design, m-learning corporations have taken on the idea of non-place as a major trope for their own work. Clark Quinn, Director of Cognitive Systems at Knowledge Planet, explains that mobile workers are big users of "non-places" and that these non-places are conducive to learning because

> You are alone. You are free from distractions. You are free from interruptions. You are free from the tyranny of meetings. These are good conditions for learning. But there is one condition that makes a nonplace even better than a library—you have no escape. You cannot jump from the plane, train or car. (Shepard, 2001, ¶ 8)

Non-places denote two conditions: always being available for production and being nowhere of significance. What is particularly problematic, however, is that Augé's non-place becomes even more pervasive when we consider electronic—not just transportation—channels and navigation. Through wireless networks we become virtual travelers, continually in motion as the network moves us through a series of communication and computation channels whether by cell phones, wireless laptops, PDAs, Radio Frequency Identification (RFID) tags, Global Positioning Systems (GPS), or other devices.

As we become communicators in transit(ion) we must remain mindful of the practical and theoretical shifts in our writing practices and contexts for composing. Indeed, our literate practices are shifting as we move our bodies and texts across spaces, making the way we draft and read fundamentally different. That is, wireless and mobile technologies influence our research collection, planning, drafting, revising, transmitting, and all other negotiations of rhetorical acts and composition practices. We understand this shift both personally and professionally. Using our wireless and mobile devices, we create, deliver, collect, read, transmit, and repurpose documents which were partially written across places—our homes, offices, planes, cars, trains, hotel rooms, coffee shops, and even lobbies. Pointing out these practices is to suggest that the very assumptions we have about our own roles as rhetoricians and writers is changed by both the discourses of and practices in the wireless and mobile technology milieu. We must consider these changes not merely as technological but also spatial, political, cultural, and social. I am advocating, then, that we be involved in articulation and rearticulation of these technologies so that a range of critical practices for wireless and mobile technologies can be reflected upon and enacted. This involvement should not be a single, inescapable way of engaging wireless and mobile work. Rather we must strive for complex, dynamic praxes of teaching and research with, through, and about wireless and mobile technologies, and we must come to view those technologies as both enabling and constraining certain aspects of our lives.

LOCAL AND GLOBAL WIRELESS INITIATIVES

The principles of non-place theory are further expressed in local and global wireless initiatives. In fact, I have purposefully selected local and global examples to illustrate the ways ubiquity is co-opted without impediment across a range of social, cultural, and geographic contexts. The two wireless programs are Empire High's 1:1 laptop-student plan and MIT's OLPC group. Although these two programs espouse to serve different communities—one the economically privileged students of Tucson, Arizona's Vail School District and the other children of "developing countries" such as

Cambodia, China, India, Brazil, Argentina, Egypt, Nigeria, and Thailand—both posit ubiquitous computing as resolving—rather than complicating—issues of engagement, access, and difference. Both the Vail and OLPC projects create an educational culture that unproblematically posits students as spontaneous learners able to internalize educational standards, pedagogues as on-demand teachers and invisible managers, and literacy as natural and pervasive. These assumptions must be not only critiqued but also challenged. Through an exploration of the discourses of these two wireless initiatives, I hope to reveal some of the key concerns we rhetoricians and compositionists may need to address as we consider the politics, practices, and policies of teaching and learning through mobile technologies such as wireless laptops.

Empire High: Vail School District's Move to Eliminate Textbooks

In July 2004, Tucson, Arizona's Vail School District hit the Associated Press wire with Superintendent Calvin Baker's announcement that newly built Empire High would eliminate textbooks and opt for laptops instead. This news quickly spread from articles in the local Tucson newspapers to national coverage in a range of outlets from CNN and the *New York Times* to *PC Magazine* and *Wired*. As both a resident of Tucson and the Vail District, I was intrigued by Superintendent Baker's announcement. Constructed practically through the integration of wireless technologies and discursively through news stories, resident bulletins, school district minutes, and the school's home page, Empire High's wireless practices forward an uncritical perspective on technology and literacy and advocate reconfigured roles for students and teachers.

Daniel Scarpinato's July 10, 2005 *Arizona Daily Star* report became a key article in the story of Vail School District's move to laptops as the primary literacy medium of its new school, Empire High. In this widely circulated article, technology is afforded a great deal of agency, particularly the ability to remediate student learning and literacy. Scarpinato (2005) explains that of the many motivating factors fueling this program, "educators are looking beyond just ratios [of student to computer]. They're looking to alter the classroom for the first time in decades by letting technology drive learning" (¶ 11). Here, wireless laptops are argued to be *the* way to foster learning, assuming that simply adding laptops will necessarily create an active learning environment. Superintendent Baker echoes this view arguing that "[w]e don't really know for sure how well the plan will work . . . I'm sure there are going to be some adjustments. But we visited other schools using laptops. And at the schools with laptops, students were just more engaged than at non-laptop schools" (Scarpinato, 2005, ¶ 26). Baker endorses the plan by suggesting students will be invested in the program, but he does not explain in this article, or any of the dozens of others published with this same quote, how many institutions were visited, when, for how long, to what extent students and teachers were prepared for the visits, or even how engagement was measured.

Interestingly, Baker's reference to "some adjustments" in his discussion of Vail's laptop program belies his own ringing endorsement of the program to fellow Arizona administrators and educators in his essay, "Time To Trash Textbooks," from the Winter 2005 *Technology in Education Newsletter*. In this article, one with a more limited readership, Baker (2005) advocates less equivocally for his district's move to 1:1 computing stating "[f]or the sake of leading instruction and learning to the place it belongs, it is time to force a radical change in line with today's work world. Textbooks need to go. Laptops need to come" (p. 3). Baker explains that laptops (the "new tools") will remediate the ills of textbooks (the "old tools") and bring us to new educational growth which is equated to being consistent with "today's work world" (Baker, 2005, p. 3). According to this view, Empire High students must have laptops, rather than a critical literacy, to be engaged, and their engagement will not guarantee self-reflective, problem-solving citizens but rather productive workers in today's economy.

The idea of more engaged students is difficult to argue against—we all want engaged learners in our classrooms—but we must question whether or not students are *automatically* engaged in their own learning by the mere presence of wireless laptops or any other technology—new or old. Wireless laptops alone cannot guarantee critical literacy, student and teacher agency, or even educational reform. If they could, even Baker might be less inclined to cast it as a matter of "force." Not surprisingly, Vail District understood the need to guarantee that the students of Empire High would make adequate use of the laptops. The district's own provisions to encourage student engagement, however, were not reported as part of the media fervor. Instead, the neighboring school district in Chandler, Arizona, reports in its January 2006 School District Minutes that "to encourage students not to loose [sic] their laptops, they [Vail district] allow students to download their own music. Insurance is also offered to parents" (Auxier, 2006, ¶ 29). Baker's unqualified praise of the laptop program to administrators and teachers, then, does not reflect either the music download enticement or the necessary insurance plan to calm parents' nervousness about laptop theft or loss. These practices demonstrate that technology is not agentless and that technology *as* engagement is a manufactured result not a natural phenomenon.

Teacher subjectivities also are constructed by the discourses and practices of Vail's 1:1 laptop program. On the one hand, the program is supposed to unleash instructor innovation through an uninterrupted stream of information to share with students; on the other, teachers often are cast as supervisors rather than active participants in learning. The *Edutopia*'s article, "No More Books: PCs Replace Textbooks at One Forward-Thinking School," explains:

> Nevertheless, Baker sees too much virtue in Empire's program for it not to spread. "If you talk to teachers, they tend to talk about the continual confinement of their profession because of No Child Left Behind and continual testing—it's like a straitjacket," he says. "Well, allowing [teachers] to use digital media instead of a prescribed textbook is opening up all kinds of creativity for them, and empowering them to do all kinds of instruction." (Colin, 2005, ¶ 9)

Rather than fighting against the "confinement" of No Child Left Behind and Arizona's Instrument to Measure Standards (AIMS) testing, which is alluded to as "continual testing" in Baker's statement, Baker redefines teacher innovation *as* technology integration. Unwired is unshackled in Baker's equation as he assumes that technology will drive improvement. This statement implies that regardless of oppressive state and federal educational mandates, teachers at Empire High should feel free, open, and ready to create new learning opportunities for students. This position of denying broader educational restrictions assumes that the wireless technologies have no material or political contexts or that the technology can transcend such restrictions. This quote is also a revision of Baker's early position that "Today, with the AIMS test, it's not the textbook that's the curriculum, it's the state standards. . . . We're getting teachers away from the habit of marching through a textbook" (Scarpinato, 2005, ¶ 26). Here again, state standards are an overarching concern for Baker, but he challenges the teachers' approach to meeting them rather than the standards themselves. Teacher agency is dually co-opted first by the imposed state and national standards and then by the district in its attempts to break their bad "habit" or free them from uninspired teaching.

The construction of literate practices and teacher subjectivities of Vail's laptop program cannot be easily separated from student subject positions. In one of the more telling reports on the 1:1 program, Dan Sorenson (2004) quotes Baker as stating "[b]ut rather than being loaded with digital versions of textbooks, the computers would have an Internet browsing program to let students access the best method available to teach skills specified by Arizona education standards" (¶ 3). Here, students are teaching themselves the standards, and teachers are absent altogether from the learning practices of technology. We educators want students to motivate their own learning, but here that learning is only related, again, to testing mandates.

While student learners are expected to control their own learning through the internalization of standards, they are also expected to police themselves in relationship to sanctioned laptop use. In Kevin Smith's (2005) report on the Vail's laptop initiative, Cindy Lee, Vail's Principal, stresses that "school administration has the ability to view student computers, including Internet history, from a remote location. Like a locker, the computers can be searched at any time . . . the school will have standards and consequences for inappropriate material found on laptops" (¶ 35-36). Students should be wary of the path they take to pursue their own education. The comparison to locker searches is interesting, because locker searches are most often associated with drugs, guns, and other contraband, but here, the laptop contraband is information.

OLPC: Bringing Laptops to Children in Developing Countries

Global efforts to construct wireless environments also abound in the language of ubiquity. OLPC is a nonprofit "dedicated to research to develop a $100 laptop—a technology that could revolutionize how we educate the world's children" (OLPC, n.d., ¶ 1). Also known as the "$100 laptop" group, this nonprofit was established through the sponsorship of faculty from MIT's Media Lab and private and corporate donors, including Advanced Micro Devices (AMD), Brightstar, Google, News Corporation, Nortel, and Red Hat. Of those leading the project, Nicolas Negroponte, author of *Being Digital* and cofounder of the MIT's Media Lab, is likely the most well known. In his famous technology tome, *Being Digital*, Negroponte (1995) envisions a world where technologies intuitively serve users' needs. He describes his ideal for the human-computer interaction as "that of a well-trained English butler. The 'agent' answers the phone, recognizes the callers, disturbs you when appropriate, and may even tell a white lie on your behalf" (p. 149). The OLPC is argued to embody the visions of Negroponte's *Being Digital*, and indeed, the OLPC project advocates for ubiquity and effectively elides many of its own class-based suppositions for the cause of technology distribution.

Making its debut at the January 2005 World Economic Forum in Switzerland, the prototype $100 laptop was unveiled with much the same media frenzy as Empire High's 1:1 program. Dubbed the "green machine" because of its bright green color (see Figure 10.1), the Linux-based prototype is outfitted "with a dual-mode display—both a full-color, transmissive DVD mode, and a second display option that is black and white reflective and sunlight-readable at 3x the resolution . . . a 500MHz processor and 128MB of DRAM, with 500MB of Flash memory . . . four USB ports . . . wireless broadband" (OLPC, 2006, ¶ 1).

Figure 10.1. OLPC'S "Green Machine" Prototype $100 Laptop

The laptop does not have a hard disk, making it incapable of high capacity storage, but it does have "innovative power (including wind-up)" (OLPC, 2006, ¶ 1). In both the design of and discourses on the $100 laptop, there is a drive to make all places, even the most economically deprived, part of the ubiquitous world of wireless and mobile technologies. The thoughtful laptop design purposefully privileges connectivity and portability over data storage needs. In fact, the FAQ page of the OLPC explains why a laptop versus a desktop:

> Desktops are cheaper, but mobility is important, especially with regard to taking the computer home at night. Kids in the developing world need the newest technology, especially really rugged hardware and innovative software. Recent work with schools in Maine has shown the huge value of using a laptop across all of one's studies, as well as for play. Bringing the laptop home engages the family. In one Cambodian village where we have been working, there is no electricity, thus the laptop is, among other things, the brightest light source in the home. (OLPC, 2006, ¶ 3)

By making the claim that the needs of children in Maine are the same as the needs of children in developing countries such as Cambodia, the OLPC erases issues of difference to insist on the need for mobile technologies. All political, economic, social, cultural, language, or other differences between children in these two contexts are eradicated as a way to privilege technology. Like Superintendent Baker, members of the OLPC believe that technology alone drives learning and engagement. Problematics such as a Cambodian estimated per capita income of $310.00 U.S. (World Bank, 2004, p. 20) all but dissolve by giving children a laptop where they can write their own software programs.

In addition to the belief that technology will erase social inequities and automatically lead to productive learning, the OLPC's program also characterizes teacher subjectivities as limited, if not entirely absent. In an October 2005 e-mail interview with *Technology Review*, Negroponte reveals his view of technology education, one that is devoid of teachers:

> Jason Pontin: I am a little skeptical about the Hundred Dollar Laptop (HDL), although I applaud the desire to improve opportunities for children in the poor world. Is technology really the main "bottleneck" (as you have said) for education in the poor world? Nicholas Negroponte: "Bottleneck" may not be the right word. Technology is the only means to educate children in the developing world. (¶ 2-3)

What is most striking across the more than 50 articles, interviews, and other resources on OLPC's own website is the absence of teachers. I could only find teachers explicitly mentioned once, and that was in Ethan Zuckerman's November 3, 2005 Blog entry, "One Laptop Per Child—a Preview of the Hundred Dollar Laptop." After noting that he "bumped into" Negroponte—a former technology advisor of his—Zuckerman explains "[w]hen I expressed some skepticism about teachers' willingness to use the computers in the classroom, he referenced Maine governor Angus King's initiative to bring computers into middle school classrooms throughout the state. Initially unpopular with teachers, the laptop project is now widely viewed as a success and is being replicated in other states" (¶ 22). Even though both Pontin and Zuckerman openly acknowledge their own skepticism, Negroponte ignores their concerns about teacher participation in the project. Another principal contributor to the OLPC, Seymour Papert, Professor Emeritus of Education and Media Technology at MIT, asserts the $100 laptop group's learning philosophy of technology "will change . . . the way children everywhere think about themselves in relation to the world. . . . It is the next big step toward . . . learning being transformed as radically as medicine, communications and entertainment" (Case, 2005, ¶ 16). Although some teachers may be part of the governmental education ministries that will be used to distribute the laptops, teachers otherwise remain an absent presence in the laptop program.

OLPC also creates student subject positions through its discourses and practices. Just as with Vail's student constructions, the OLPC defines the student recipients of its program as both self-guided, engaged learners and potential victims or perpetrators of crime. Negroponte suggests that OLPC students will have the advantage of "seamless" learning (Twist, 2005, ¶ 9), and he further stresses that "studies have shown that kids take up computers much more easily in the comfort of warm, well-lit rich country living rooms, but also in the slums and remote areas all around the developing world" (Twist, 2005, ¶ 10). Glaringly, Negroponte ignores the economic constraints brought to bear on OLPC students. These students through their technologies will transcend the economic disparities of "slums" to develop new software programs—through the open source software on their machines—and perhaps a few of the best and brightest, according to Papert, will use the machine for entrepreneurial purposes (Barylick, 2006, ¶ 16). These positions seem to ignore the place-based facts that Cambodia, a country most often cited by the OLPC, has been beset with political and economic problems including the abolition of the entire educational system from 1975-1979 during the Khmer Rouge regime. Even today, Cambodia's Ministry of Youth, Education & Sports (MYES) reports that,

> Due to budget constraints there is little hope of ensuring equal opportunity for every child to receive a 9-year basic education in the very near future. . . . Because of poverty, shortage of labor in the family and the low level of education, certain segments of the population perceive that education is not imperative for themselves nor their families. (1999, ¶ 3-5)

Cambodia is just one of the many nations targeted by OLPC, but this country's own infrastructure and economic health seem to reflect intense, yet ignored, contextual issues.

The OLPC's granting technology access to children of the developing countries often seems motivated more by a desire for technology innovation than a commitment to educational potential. In her critique of electronic colonialism, Cynthia L. Selfe (1999a) explains our roles in this phenomenon as "cybertourists and cybercapitalists who both understand and represent the world as a private standing reserve" (p. 299). These roles are aptly demonstrated in one of OLPC's widely distributed promotional images for the "green machine." Figure 10.2 depicts the green machine's screen where the smiling faces of potential OLPC students are "captured." The background of the entire image is white—devoid of place, a context-free environment. As the viewer of this image, we assume the role of the white child interlocutor controlling the machine and the inhabitants of color in it. The OLPC project relies on confining student subjectivities to

Figure 10.2. Green Machine Promotional Image

meet and enact its philosophies on technology and education, making the recipients of the laptops "private standing reserve" happy to be part of the program.

In terms of their responsibility for maintaining the security of the green machines, OLPC students are similarly hailed into the role of either crime victims or criminals. Negroponte reveals that the "green machine" is green to "discourage theft in the developing countries [where] the units will be sent to and makes the item distinctly identifiable" (Barylick, 2006, ¶ 8). Color is only one means of discouraging theft; digital security strategies are another. In this same *UPI* report by Chris Barylick (2006), Negroponte discloses that "the laptops, which communicate wirelessly with each other, can be configured to deactivate if they haven't been turned on and in contact with other units, as they would in a school environment" (¶ 10). As learners, the students are fully responsible for motivating their own learning, but as users they are seen as less reliable, possible victims or criminals. If the laptops are not enmeshed in the local wireless network, they will be ineffective, even as light sources for their poor village homes.

In both the discourses on and design of Vail's 1:1 program and OLPC's initiative, mobile and wireless technologies are defined by their assumed ubiquity. Without consideration of the spatial, historic, economic, language, or other cultural barriers, students and teachers are responsible for ensuring the unmitigated success of such programs, ones that equate technology access with engaged learning and literacy (see Turnley in this collection and Moran, 1999, and Grabill, 2003, for further discussions on technological access). These programs stress that students are to motivate their own learning and teachers need to either stay out of the way of student learning or police student online behavior. Although I applaud efforts to increase critical literacy and open up new possibilities for reconfiguring student and teacher agency, these programs and our own cannot achieve such laudable goals without paying careful attention to the context of technology teaching and learning, without situating these technologies within particular spatial configurations, and without rearticulating broader, more productive roles for students, teachers, and administrators.

Starting Points to Critical Mobile and Wireless Technology Integration

From discussions of Wi-Fi marketing (Moeller, 2004) to constructions of mobile and wireless discourses and pedagogies (Dean, Hochman, Hood, & McEachern, 2004; Graham Meeks, 2004; Zoetewey, 2004), scholars in our field are beginning to examine the impact of mobile and wireless technologies on our programs and classrooms. As more campuses move to wireless networks, we WPAs and composition teachers are likely to find ourselves using, discussing, and making policies for a range of mobile and wireless devices. Cultural assumptions about ubiquity, however, present challenges to the ways we create and enact mobile and wireless pedagogies. Programs such as Vail's and OLPC's make individual student, teacher, and administrator needs seem placeless, unbound by material conditions. Rather than characterizing literacy, agency, and student and teacher roles in deterministic ways, we must use our decision-making positions to re-articulate mobile and wireless technologies around critical literacy practices and renegotiated student and teacher subject positions. Unlike placeless articulations of ubiquity represented in Empire and OLPC projects, our approaches must carefully attend to the social, institutional, political, and cultural forces shaping our local contexts. This closing offers just a few starting points to re-articulate grounded, critical curricular and pedagogical practices for mobile and wireless composing.

For WPAs, administrative considerations of mobile and wireless integration must include attention to workload, security, and privacy issues. Hyped by wireless initiatives, perpetual contact between students and teachers creates an expectation of constant, continual engagement. This anytime, anywhere position carries risks of exploitation, especially for part-time, nontenure–track faculty, and working students. Security and privacy issues also must be addressed to help both teachers and students negotiate the new roles of being mobile teachers and learners. A WPA can concentrate on these workload, security, and privacy concerns through open meetings, discussion forums, and standing advisory committees. These conversations must involve, or at least represent, key stakeholders such as students, graduate students, instructors, staff, and faculty in negotiating expectations for student-teacher and student-student interaction. Setting boundaries for electronic office hours, establishing guidelines for online and face-to-face student-teacher interaction, and combating the assumption that teachers or students must be available for immediate electronic communication 24/7 may be necessary in creating a balance in the workload of wireless pedagogies.

WPAs also can collaboratively construct wireless and mobile policies and curricula that strive to acknowledge, not erase, issues of differential access and experience. Thus, WPAs must work to bring collaborative and careful decision making to bear on the newly configured spaces of mobile and wireless pedagogies (see Brooks and Poe & Garfinkel in this collection for suggestions about administrative concerns). The WPA must advocate for wireless resources for all members of his or her program, especially if the assumption is that laptops are the new norm established through campus wireless programs. Pushing all costs and consequences on to either teachers or students can result in further disparities and may silence, rather than include, a range of social and cultural perspectives.

Teachers themselves can work to assert their own subjectivities within wireless policies and pedagogies. Perhaps more than any other faculty, those of us teaching composition are acutely aware of the challenges to retain students, support their intellectual growth, and guide them in the development of critical literacy. We teachers should seek out ways to contribute to the broader campus initiatives to integrate wireless—sitting on committees, fostering collaborations with IT departments, and seeking out grants and other sources of support. All of these efforts can help to challenge the rigid subject positions of on-demand teacher or computer police officer constructed by deterministic wireless programs. Further, we can enact classroom pedagogies that include critical and rhetorical examinations of local campus and community mobile and wireless technology integrations. Students can investigate, map, and compose a range of texts to challenge deterministic positions on mobile and wireless technologies. These projects might include inviting students to keep a "travelogue" of their technology experiences for a number of days (verbal, visual, and auditory texts can enhance their records). They then can analyze their logs individually and collaboratively examining their modes of travel, ways in which technologies impeded or enhanced their interactions with others, and how their logs reflect certain social, cultural, and economic differences. They might even construct a collaborative visual or electronic map based on their logs, one that attempts to complicate constructions of movement and space. Students also might investigate governmental bills such as the Minority Serving Institution Digital and Wireless Technology Opportunity Act of 2005. They can examine how such bills account for technological disparities, construct race and class, and exhibit certain values. As a response, they could write their own legislation, compose letters to government officials, or construct responses to technological inequities that strive for balance. Students might even search for wireless security and privacy policies that apply to them as students, workers, or community members. They could consider the ways in which such policies make assumptions about them and their actions/inactions. They might even write or revise such policies to attend to a range of social issues they face as students, learners, and citizens. Each of these suggestions must be part of a much larger process of contextualizing mobile and wireless technologies, and all must be refined taking into consideration student and teacher needs at particular institutions. The hope is that such prompts can enable students and teachers to question and even enact situated—rather than placeless—pedagogies of mobile and wireless technologies (see Bjork & Pedro; Brown; Kitalong; and Moeller in this collection for other mobile and wireless pedagogical practices).

Instantiating a critical perspective on mobile and wireless technologies will not be an easy endeavor. The pull of ubicom determinisms is difficult to negotiate and is already circulating in programmatic discourses and practices of wireless initiatives. We composition administrators and teachers, however, must actively challenge positions that align technology integration with a simplified notion of literacy, with rigid teacher and student subjectivities, and with placelessness. Without a re-articulation of the discourses and practices on mobile and wireless devices, we may find ourselves complicit in the inequities represented by programmatic mobile and wireless initiatives.

ENDNOTES

1. In August 2001, *Wireless Internet* reported that "[m]obility-focused applications and consumer-based data services and devices will drive the North American wireless data market to grow from 7.3 million subscribers in 2000 to 137.5 million subscribers in 2005, according to Dataquest Inc., a unit of Gartner Inc." (¶ 1).
2. See research by Srivastava, Muntz, & Potkonjak, 2001 for a discussion on the development of a Smart Kindergarten.

REFERENCES

Apple Computer, Inc. (2003). Profiles in success: Henrico county public schools continuous learning. Retrieved November 26, 2005, from http://www.apple.com/education/profiles/henrico1/

Augé, Marc. (1995). *Non-places: Introduction to an anthropology of supermodernity* (John Howe, Trans.). New York: Verso.

Auxier, Annette. (2006, March 8). Minutes of a meeting of the governing board, Chandler unified school district. Retrieved March 30, 2006, from http://ww2.chandler.k12.az.us/governing-board/minutes.doc

Baker, Calvin. (2005, Winter). Time to trash textbooks. *Technology in Education Newsletter, 3.*

Barylick, Chris. (2006, February 9). One laptop project reaches critical stages. *UPI.* Retrieved February 20, 2006, from http://www.upi.com/Hi-Tech/view.php?StoryID = 20060 209-124347-1619r

Case, Christa. (2005, November 16). A low-cost laptop for every child: Effort to link the world's rural poor to the Internet with a $100 computer gets a boost from the United Nations. *The Christian Science Monitor.* Retrieved February 20, 2006, from http://www.csmonitor.com/2005/1116/p 04s01-ussc.html

Colin, Chris. (2005). No more books: PCs replace textbooks at one forward-thinking school. *Edutopia.* Retrieved February 12, 2006, from http://www. edu-topia.org/magazine/ed1article.php?id = Art_1359&issue = oct_05

Dean, Christopher, Hochman, Will, Hood, Carra, & McEachern, Robert. (2004). Fashioning the emperor's new clothes: Emerging pedagogy and practices of turning wireless laptops into classroom literacy stations. *Kairos, 9*(1). Retrieved March 20, 2005, from http://english.ttu.edu/kairos/9.1/binder2.html?cover-web/hochman_et_al/intro.html

DeCerteau, Michel. (1984). *The practice of everyday life* (Steven Rendall, Trans.). Los Angeles: University of California Press. (Original work published 1980)

Dertouzos, Michael. (1999, July). The future of computing. *Scientific American, 281,* 52-56.

Foucault, Michel. (1978). *Discipline and punish: The birth of the prison* (Alan Sheridan, Trans.). New York: Pantheon. (Original work published 1975)

Foucault, Michel. (1980). Questions on geography. In Colin Gordon (Ed.), *Power/knowledge: Selected interviews and other writings 1972-1977* (pp. 63-77). New York: Pantheon. (Original work published 1976)

Foucault, Michel. (1984). Space, knowledge, and power. In Paul Rabinow (Ed.), *The Foucault reader* (pp. 239-256). New York: Pantheon Books. (Original work published 1982)

Foucault, Michel. (1997). Of other spaces: Utopias and heterotopias. In Neil Leach (Ed.), *Rethinking architecture: A reader in cultural theory* (pp. 350-356). New York: Routledge. (Original work published 1967)

Gateway Computers. (2004). Enterprise for higher education: Reliable answers for a robust campus network [Brochure]. Irvine, CA.

Grabill, Jeffrey T. (2003). On divides and interfaces: Access, class, and computers. *Computers and Composition, 20*, 455-472.

Hawisher, Gail E., & Selfe, Cynthia L. (1991). The rhetoric of technology and the electronic writing class. *College Composition and Communication, 42*, 55-65.

IBM. (2003). IBM ThinkPad university program: Education on demand [DVD]. New York: International Business Machines Corporation.

Johnson-Eilola, Johndan. (1997). *Nostalgic angels: Rearticulating hypertext writing.* Norwood, NJ: Ablex.

Johnson-Eilola, Johndan. (2005). *Datacloud: Toward a new theory of online work.* Cresskill, NJ: Hampton Press.

Kaplan, Nancy. (1991). Ideology, technology, and the future of writing instruction. In Gail E. Hawisher & Cynthia L. Selfe (Eds.), *Evolving perspectives on computers and composition studies: Questions for the 1990s* (pp. 11-42). Urbana, IL: NCTE and Computers and Composition Press.

Kimme Hea, Amy. (2008). Riding the wave: Articulating a critical methodology for web research practices. In Heidi A. McKee & Dánielle Nicole DeVoss (Eds.), *Digital writing research: Technologies, methodologies, and ethical issues* (pp. 269-286). Cresskill, NJ: Hampton Press.

Kingdom of Cambodia. (1999). Ministry of education, youth and sports access and participation. Retrieved, February 20, 2006, from http://www.moeys.gov.kh/profile/ edu_in_cambodia/access_participant.htm

Lefebvre, Henri. (1991). *The production of space* (Donald Nicholson-Smith, Trans.). Malden, MA: Blackwell. (Original work published 1974)

Manjoo, Farhad. (2001, March 13). Technology: Is that all there is? *Wired.* Retrieved February 17, 2006, from http://www.wired.com/techbiz/media/news/2001/03/41971

Mauk, Johnathan. (2003). Location, location, location: The "real" (e)states of being, writing, and thinking in composition. *College English, 65*, 368-388.

Meeks, Melissa Graham. (2004). Wireless laptop classrooms: Sketching social and material spaces. *Kairos, 9*(1). Retrieved January 4, 2006, from http://english.ttu.edu/kairos/9.1/binder2.html?coverweb/meeks/index.html

Moeller, Ryan. (2004). Wi-fi rhetoric: Driving mobile technologies. *Kairos, 9*(1). Retrieved January 4, 2006, from http://english.ttu.edu/kairos/9.1/binder2.html?coverweb/moeller/index.html

Moran, Charles. (1999). Access—The "A" word in technology studies. In Gail Hawisher & Cynthia Selfe (Eds.), *Passions, pedagogies, and 21st century technologies* (pp. 205-220). Logan: Utah State University Press.

Mountford, Roxanne. (2001). On gender and rhetorical space. *Rhetoric Society Quarterly, 31*, 41-71.

Negroponte, Nicholas. (1995). *Being digital.* New York: Alfred A. Knopf.

OLPC. (2006). Frequently asked questions. Retrieved March 6, 2006, from http://laptop. org/FAQ.html

OLPC. (n.d.). One laptop per child. Retrieved February 5, 2005, from http://laptop.org

Payne, Darin. (2005). English studies in Levittown: Rhetorics of space and technology in course-management software. *College English, 67*, 383-408.

Pontin, Jason. (2005, October 13). The hundred dollar man: *Technology Review's* editor in chief talks with Nicholas Negroponte about the hundred dollar computer. Retrieved February 17, 2006, from http://www.technologyreview.com/InfoTech/ wtr_14874,294,p1.html

Porter, James E., & Sullivan, Patricia. (1997). *Opening spaces: Writing technologies and critical research practices.* Greenwich, CT: Ablex.

Porter, James E., Blythe, Stuart, Grabill, Jeffrey T., & Miles, Libby. (2003). Institutional critique revisited. *Works and Days, 41/42*, 219-237.

Porter, James E., Sullivan, Patricia, Blythe, Stuart, Grabill, Jeffrey T., & Miles, Libby. (2000). Institutional critique: A rhetorical methodology for change. *College Composition and Communication, 51*, 610-642.

Reynolds, Nedra. (1998). Composition's imagined geographies: The politics of space in the frontier, city, and cyberspace. *College Composition and Communication, 50*, 12-35.

Reynolds, Nedra. (2004). *Geographies of writing: Inhabiting places and encountering difference.* Carbondale: Southern Illinois University Press.

Scarpinato, Daniel. (2005, July 10). All-laptop high school to open in Vail. *Arizona Daily Star.* Retrieved January 30, 2005, from http://www.azstarnet.com/dailystar/dailystar/83469

Selfe, Cynthia L. (1999a). Lest we think the revolution is a revolution: Images of technology and the nature of change. In Gail E. Hawisher & Cynthia L. Selfe (Eds.), *Passions, pedagogies, and 21st century technologies* (pp. 292-322). Logan: Utah State University Press.

Selfe, Cynthia L. (1999b). Technology and literacy: A story about the perils of not paying attention. *College Composition and Communication, 50*, 411-436.

Selfe, Cynthia L. (1999c). *Technology and literacy in the twenty-first century: The importance of paying attention.* Carbondale: Southern Illinois University Press.

Selfe, Cynthia L., & Selfe, Richard J. Jr. (1994). The politics of the interface: Power and its exercise in electronic contact zones. *College Composition and Communication, 45*, 480-504.

Shepherd, Clive. (2001). M is for maybe. *Fastrak Consulting Ltd.* Retrieved February 13, 2006, from http://www.fastrak-consulting.co.uk/tactix/features/mlearning.htm

Smith, Kevin. (2005, July 14). New Vail schools built to fill niches: Technology, population the driving force behind two projects. *Arizona Daily Star.* Retrieved January 30, 2006, from http://www.azstarnet.com/dailystar/relatedarticles/83956.php

Soja, Edward W. (1989). *Postmodern geographies: The reassertion of space in critical social theory.* London: Verso.

Soja, Edward W. (1996). *Third space: Journeys to Los Angeles and other real-and-imagined places.* Oxford, UK: Blackwell.

Sorenson, Dan. (2004, March 18). Vail school may shun textbooks for laptops. *Arizona Daily Star.* Retrieved March 9, 2005, from http:// www. azstarnet.com/daily star/allheadlines/14295.php

Srivastava, Mani, Muntz, Richard, & Potkonjak, Miodrag (2001). Smart kindergarten: Sensor-based wireless networks for smart developmental problem-solving environments. *Proceedings of the ACM SIGMOBILE Conference,* pp. 132-138.

Sullivan, Patricia A., & James E. Porter. (1993). Remapping curricular geography: Professional writing in/and English. *Journal of Business and Technical Communication, 7,* 389-422.

Twist, Jo. (2005, November 17). UN debut for $100 laptop for poor. *BBC News.* Retrieved February 20, 2006, from http://news.bbc.co.uk/1/hi/technology/4445060. stm

U.S. Department of Education. (2004). Toward a new golden age in American education—How the internet, the law and today's students are revolutionizing expectations. *National Education Technology Plan 2004.* Retrieved November 26, 2005, from http://www.ed.gov/about/offices/list/os/technology/plan/2004/plan_pg10.html#success

Virilio, Paul. (1986). *Speed and politics: An essay on dromology* (Mark Polizzotti, Trans.). New York: Semiotext(e). (Original work published 1977)

Weiser, Mark. (1991, September). The computer for the 21st century. *Scientific American, 265,* 94-104.

Weiser, Mark, & Seely Brown, John. (1996). The coming age of calm technology. Xerox PARC. Retrieved February 20, 2006, from http://www.ubiq.com/hypertext/ weiser/acmfuture2endnote.htm

Wireless Internet. (2001, August). Dataquest says there will be 137 million wireless data users in North America by 2005. Retrieved March 9, 2005, from http://findarticles.com/p/articles/mi_m0IGV/is_8_3/ai_78359169

World Bank. (2004). World bank annual report. Retrieved March 6, 2006, from http:// www.worldbank.org/annualreport/2004/Vol_2/PDF/WB%20Annual%20Report%202004.pdf

Zoetewey, Meredith. (2004). Disrupting the computer lab(oratory): Names, metaphors, and the wireless writing classroom. *Kairos, 9*(1). Retrieved January 4, 2006, from http://english.ttu.edu/kairos/9.1/binder2.html?coverweb/zoetewey/index.html

11

WRITING IN THE WILD

A PARADIGM FOR MOBILE COMPOSITION

Olin Bjork

John Pedro Schwartz

In *English Composition as a Happening*, Geoffrey Sirc (2002) compares the writing classroom to a museum where the instructor guides students on a tour of the "Great Works" in composition readers and offers such texts as models for student imitation. Drawing on avant-garde concepts from the 1960s, Sirc calls for an "*other scene* of writing instruction" akin to a "museum without walls" where the "outside is let inside" (pp. 1, 294, 286).[1] By bringing into the classroom atmospheric objects (e.g., candles) as well as new objects of study (e.g., rap music), Sirc hopes to foster new habits of thought and enliven student writing. Despite this change of scene, teaching still "happens" in the classroom and students continue to write in conventional spaces (e.g., the library or dorm room). As classrooms, campuses, and communities go wired or even wireless, the computer increasingly functions as a portal, letting the outside world in. Although students are now more likely to write in off-campus spaces (e.g., a coffee shop), they are less inclined to do research anywhere but on the Internet. These conventional spaces of research and writing tend, however, to be physically remote from many of the topics students write about.

To combat the notion of writing as an isolated activity, some composition instructors promote fieldwork. In *The Call to Write*, John Trimbur (2001) explains that field research is sometimes necessary because certain questions "can't be addressed solely on the basis of print or electronic sources" (p. 549). This rationale implies that if all questions could be answered through secondary sources, there would be no need for students to do fieldwork. Other composition scholars, such as Sidney I. Dobrin and Christian R. Weisser (2002), support fieldwork less from the standpoint of justifying primary research than from the perspective of public service, experience, and ecological awareness: "Nature and environment must be lived in, experienced to see how the very discourses in which we live react to and with those environments" (p. 57). In the vast majority of fieldwork models, however, the writer returns to a conventional writing space—library, dorm room, or other such location—after collecting data at the site of the object. Few instructors require that all acts of composition—research, writing, and publication—take place *in* the field.

We propose a paradigm for mobile composition in which students visit places of rhetorical activity (e.g., city parks, waiting rooms, shopping malls) and research, write, and (ideally) publish on location. By publish, we mean transmit their writing to their target audience. Although the current proliferation of wireless devices and networks facilitates mobile composition, researching and writing at the site of the object (the place of rhetorical activity) is also possible with other (older) mobile technologies, such as pen and notepad. Assignments that require students to compose *in situ* using mobile technologies help them achieve insight into the relationship between discourse and place that Nedra Reynolds (2004) calls "inhabitance" (p. 4). Moreover, such assignments aid students in recognizing what Bruce Horner (2000) calls the "materiality of writing." For Horner, this term refers to the material structures determining writing production, which range from communication technologies to global relations of power, from particular subjectivities to socioeconomic formations, from physical classroom settings to organizational structures (Horner, 2000, pp. xviii-xix). Anne Frances Wysocki (2004) builds on Horner's notion of the materiality of writing as it exists for teachers and students in composition classrooms. She defines new media texts not as digital or multimodal but as texts that draw attention to their materiality, and she recommends viewing them as objects rather than as immaterial, invisible carriers of meaning (p. 22). According to Wysocki, students aware of writing as a material social practice recognize that what they write with determines what they write, as well as "see—through what they write—their particular locations in time and place, and hence how they are shaped by but can in turn shape those locations (and themselves) through textual work" (p. 4).

Although composition scholars such as Horner and Wysocki have rejected idealized notions of writing as an individual practice separate from material circumstances, they tend to maintain a sense of writing as an activity that occurs in a conventional space once a student attains awareness of these determining circumstances. In emphasizing social, cultural, and historical contexts, Horner and Wysocki overlook the importance of the physical contexts of "textual work," forgetting that *where* students write determines not only *what* they write but also what they write *with*. Similarly, instructors who teach writing as a cultural, situated act often craft assignments that presuppose a clean, well-lighted writing space such as the library or dorm room. The problem is that these spaces homogenize the same material differences that instructors are trying to underscore. For example, an African American student writing at home is far more disembodied than if she were writing in the visible, public space of an art museum. Similarly, an affluent female student writing at a working-class, male-dominated bowling alley feels her status more acutely than she would feel it writing in the library or the dorm room. We argue that students can better perceive—and learn to challenge—their social, cultural, and historical locations when they research, write, and even publish *on location*. Mobile composition helps students understand the interdependency of agency and material structures by confronting students with the effects of these structures and by connecting writing assignments to spaces and technologies. When using a tablet computer to document social interactions at an event, for example, a student may begin to see that writing assignments, spaces, and technologies are mutually determining. This insight may lead to the further realization that the material conditions shaping what students write and who they become through writing are fluid and changeable. Mobile composition also facilitates at least two other pedagogical objectives: Teachers of new media writing can ask students to venture into the field and use multimedia devices to produce combinations of text, image, audio, and video, and instructors interested in experiential education can encourage students to interact with a wider range of environments.

In the section that follows, we discuss how wired and wireless technologies have reconfigured the space of writing instruction. Then, we examine how even the broadest notions of the materiality of writing are complicated by increased mobility, which encourages writing processes and genres that depart from the essay drafts common to both homework and fieldwork models of composing. Finally, we look at three instances of mobile composition to suggest pedagogies that relocate composition in the field, which, because it is untamed for writing, we will call "the wild."

WIRED AND WIRELESS CLASSROOMS

The evolution of the writing classroom from unwired to wired to wireless has led to the emergence of a strange, hybrid entity: the laboratory/classroom. Whereas some wired classrooms only offer one computer with an Internet connection—inevitably monopolized by instructors—computer labs/classrooms usually provide each student with a networked computer. This configuration blurs the material distinctions between spaces of instruction and spaces of composition, positioning instructors and students as co-experimenters working in a productive lab environment. The problem with this model is that it further privileges the classroom space, which now houses cutting-edge writing technologies. Teachers and students become dependent on machines located in a particular site to which they often have limited access.

The days of the laboratory/ classroom model may well be numbered. Two of the four CoverWeb articles on "The Rhetoric and Pedagogy of Portable Technologies" in the fall 2004 issue of *Kairos* discuss the "wireless laptop classroom." This rubric describes two fundamentally different models of hardware and software dispensation, revealing that the adjectives *wireless* and *portable* (mobile) are far from synonymous. In the cart model, students are granted access to a laptop with wireless connectivity and software for the duration of the class and only within the walls of the classroom. In the student-owned model, students are required to bring a laptop with wireless connectivity and software to the classroom. The main problem with the cart model is inscribed in the title of Christopher Dean, Will Hochman, Carra Hood, and Robert McEachern's article, "Fashioning the Emperor's New Clothes: Emerging Pedagogy and Practices of Turning Wireless Laptops into Classroom Literacy Stations" (2004). To be conceived as "classroom literacy stations," laptops must be either physically stationary or constrained by the walls of a classroom or classroom building. Thus, the emperor is exposed as the old wired laboratory/classroom in new wireless garb. At the authors' institution, Southern Connecticut State University, laptops are delivered to students via a cart, specifically an "iBook Wireless Mobile Lab," which charges them and includes a wireless router and printer. Students use the laptops wherever they choose within the classroom, but then the students must return the laptops to the cart at the end of the class period. At the Computer Writing and Research Lab of the University of Texas, the situation is even more restrictive. The laptop classroom remains wired, with each machine tethered by power and Ethernet cords to hexagonal tables. Although the room has a wireless signal, the inflexibility of its seating, security measures, and network protocols all keep the machines largely immobile. In this static configuration, the only advantages of laptops over desktops are their small footprint, low profile, and close-ability.

While the cart model levels the hardware and software playing field if not the computing skills of the students, the student-owned model makes students responsible for the use and maintenance of their own machines, which often vary widely in speed, memory, and power. In "Wireless Laptop Classrooms: Sketching Social and Material Spaces," Melissa Graham Meeks (2004) of the University of North Carolina seeks to minimize this problematic of ownership: "Already marked by race, gender, class, age, able-bodiedness, and region-of-origin among a myriad other differences, students need not also be marked by the functionality of their laptops" (Marked by, ¶ 2). Assuming this inequality can be mitigated through a standard laptop program, the student-owned model is ultimately preferable to the cart model because it makes the classroom space less rarefied and narrows the technical knowledge gap between students and instructors. But instructors who anticipate a move to the student-owned model have other questions to ponder: Does the laboratory label still apply? How can the computer-assisted writing classroom maintain its new-found position of privilege in the academy when it becomes the pedagogical equivalent of any other enclosed space that offers seating and a Wi-Fi hotspot?[2]

This dilemma of spatial status is underscored by Meredith Zoetewey (2004) in "Disrupting the Computer Lab(oratory): Names, Metaphors, and the Wireless Writing Classroom." Zoetewey teaches at Purdue University, where the lab metaphor was first used in 1976 for its Writing Lab. (Similar tutoring places at other institutions are often named "writing centers.") In 1994, the Purdue Writing Lab developed a supplemental website, the O(nline) W(riting) L(ab). Soon, other writing programs adopted both the name and the concept. As Zoetewey attests, English departments recognize that the lab metaphor links their disciplines to the paradigm of scientific knowledge–making preferred by the university: "At Research 1-designated institutions, occupying a virtual or physical lab can lend cachet to the goings on of an already dubious enterprise—English studies" (Labs, ¶ 1). Although the lab metaphor generates money for computers and networks, Zoetewey finds it troubling because the traditional scientific laboratory is an "arhetorical, ersatz place" (A paradox, ¶ 2) in which women and other disenfranchised people were only recently granted full participation (Modest witnesses, ¶ 3). The metaphor seems apt when the room houses hardware and software, the respective analogues to experimental apparatus and chemical sets. When students bring their own hardware and software, however, the metaphor loses force.

MOBILIZING COMPOSITION

As students, classrooms, and campuses go wireless, composition instructors may seek to develop pedagogies responsive to this trend. Because these pedagogies will have different goals, they will require different devices. The iPod, already in the hands of many students, is primarily designed for storing and playing audio files and may therefore be inappropriate for courses that make frequent use of visuals. Even iPods with photos or video display are not designed to record images. Nor are iPods necessarily well equipped for taking class notes. Professors at Duke and other institutions who have agreed to provide course materials for iPods or allow their lectures to be podcast have found that more students are skipping class and relying on their iPods to keep up (Cohen, 2005, ¶ 4). Laptops are better suited for text entry than iPods and other handhelds, but they are not as portable and generally require a flat working surface with low ambient light. Despite many advantages, laptops have failed to overtake paper as the campus note-taking medium of choice because laptops are expensive, noisy, and may create a barrier between teachers and students. Tablet computers, meanwhile, combine the qualities of laptops with those of paper products.[3] Their sketchpad capability may make them a more versatile tool for taking notes than laptops or handhelds.

Composition teachers who encourage students to participate in service-learning projects, microethnographies, documentaries, or case studies may want to begin their mobile technology exploration by researching departments that have experimented with tablets in fieldwork. At the University of Texas, for example, Microsoft donated Compaq tablets to the Department of Architecture's graduate program in Community and Regional Planning (Foster, 2003). Graduate students took these tablets to a neighborhood on the Mexican-U.S. border, where they worked to design a park. "'I couldn't believe I was standing in the middle of a field . . . in a place without many of the basic necessities, and there I was writing, actually digitizing notes on my tablet,' says Marilyn Shashoua. 'It is as if you are carrying a clipboard'" (Foster, 2003, p. A33). Handwriting and drawing are impossible with a laptop, and typing while standing is difficult. Tablet users can save their digitized stylus-written notes as images or convert them to text documents using character recognition software. Though far from flawless, this conversion process means that field research can be documented on site, in the field, in a printable form. Despite the advantages of inscription and portability, however, tablets can become nearly as difficult to carry for extended periods of time—or view in direct sunlight—as laptops.[4] Handheld computers, such as PDAs and smartphones, are smaller, lighter, and brighter than tablets or laptops, but like iPods, these devices are not designed for heavy text input. A new device that recently debuted, the handheld tablet PC, may become the ideal tool for mobile composition assignments that require handwriting and drawing. Other handhelds are better suited for assignments that emphasize images, audio, and/or video.

Although mobile composition is not yet advocated at the institutional level, distance education movements and iPod initiatives demonstrate that universities are willing to explore ways of moving instruction out of classrooms and into places where students live and study. Should composition follow this trend toward spatially distributed education? Most writing instructors would probably agree with Naomi Baron when she says, "I want to believe that what I'm doing in class is not canned and has something to do with the people who are there" (qtd. in Cohen, 2005, ¶ 27). But what teachers do "in class" also has something to do with where they do it—the classroom itself "cans" their pedagogies. Wireless networks have the potential to redefine the classroom as any physical or virtual place where instructors and students happen to meet. Nonetheless, the physical classroom continues to be a highly practical space for face-to-face composition instruction and instructor-student and student-student collaboration. Instead of mobilizing every dimension of our instruction, we may consider first mobilizing student research, writing, and even publication.

Incorporating a mobile composition assignment into a syllabus helps to productively destabilize the traditional classwork-homework dichotomy. The term *homework* implies a number of assumptions—temporal, spatial, technological—that tend to blur because of composition's focus on *process*. For example, Nedra Reynolds (2004) argues that common process metaphors such as "drafts" have obscured rather than illuminated the material dimensions of the writing process:

> Given today's writing tools, it has become increasingly silly to ask students for "drafts" that demonstrate the writing process when so much of writing takes place on the screen in a more fluid, spatial medium that doesn't lend itself well to "frozen" representations. Electronic writing technologies enfold, embed, or dilute what we have known as the

writing process, make it more invisible, harder to see—often hiding the role of invention as well as the differences among writers and their composing habits. Making acts of writing or composing processes *more* visible is the point, rather than trying to eliminate process from our language or theories. (pp. 5-6)

The "electronic writing technologies" that make writing processes "more invisible" include word processors and other WYSIWYG editors, whose very names (*processors* and *editors*) suggest an appropriation of agency and a specification of genre. Electronic essays and websites are works-in-progress—only the latest iteration appears on the screen, making it difficult to preserve a record of the writing process. But some genres and technologies do make the writing process "*more* visible." Left to their own wired or wireless devices, students are far more likely to use them to compose email, text, or instant messages and create social-networking web pages or blogs than to write essays or build websites. Some students prefer process-as-product genres such as blogs because they preserve a history of written interaction. Blogs are self-organizing, archiving periodicals. Unlike print newspapers or journals, blogs are editable after publication, but the time stamp feature of blogs often fosters an ethic of nonreplacement: Most bloggers refuse to edit already published entries, preferring instead to fine-tune their arguments through the comment feature or subsequent entries. Similarly, despite efforts by instructors to naturalize a highly conventional process of revision that removes or replaces text, students continue to resist making substantial changes to writing they have already made public. Most students would rather apply what they learn from one writing situation to the next.

As the next writing situation might be around the next corner, students equipped with mobile technologies are always prepared to research, write, and even publish on location. Although in some cases students will not have access to wireless devices and networks, mobile composition does not depend upon such technologies. A pencil and paper will also allow students to compose at the site of the object. Furthermore, the growing parity of student access to wired technologies demonstrates that digital divides narrow when institutions take the first step toward bridging them. Increasing institutional support for wireless campuses will encourage students to buy compatible devices. In addition to contending with difficult questions of technological access, mobile composition pedagogies must attend to physical and cultural access to public spaces: Some instructors may rightly worry that this approach may mark students according to their mobility. Indeed, transportation, race, gender, class, age, and disability are determining factors in real and perceived mobilities. But mobile composition can encourage students to confront such differences, for one goal of repositioning writers in the wild is to foster awareness of their social, cultural, and historical locations.[5]

FIGURING MOBILE COMPOSITION

A look at three examples of mobile composition should prove useful for instructors seeking to move student research, writing, and even publication from conventional spaces to the site of the object. The first example involves the use of smartphones to create a museum of everyday life. Through web and mobile technologies, museums today make their cultural and educational resources available to remote audiences. Reversing Sirc's (2002) scenario of a "museum without walls" where "the outside is let inside," many museums let the inside outside (pp. 294, 286). The problem, argues Konstantinos Arvanitis (2005) of the University of Leicester's Department of Museum Studies, is that this use of new media fails to create productive collaborations among virtual visitors, "real" visitors, and curators. Despite creating opportunities for visitors to explore collections online, museologists have not yet fulfilled Eilean Hooper-Greenhill's (1992) call for the "post-museum"—a place where "knowledge is constructed, rather than transmitted" (Arvanitas, 2005, p. 3). As part of his doctoral research, Arvanitis asked ten smartphone users to compose Multimedia Messaging Service (MMS) image-texts about three archaeological monuments in Thessaloniki, Greece.[6] The goal of the study was to understand how visitors make meaning out of their casual encounters with ancient monuments. More broadly, Arvanitis' study (2005) aimed "to bring the voices [and objects] of the everyday into the museum" (p. 3) in order to create the experience of a "museum *outside* walls" (p. 7). The MMS messages revealed the extent to which the monuments had become integrated into the scenery of everyday life. For example, values such as pedestrian convenience and political statement often trumped concern for the monuments' cultural and historical significance.

By documenting and interpreting objects for exhibition, the participants in Arvanitis' study assumed the role of curators. Yet their curator status posed an alternative to the traditional institutional discourse of museum exhibits in several important respects. Whereas a conventional curator's captions, wall texts, brochures, and catalogues focus on exhibited objects, the alternative or everyday curators' MMS image-texts depicted unplanned interactions with the monuments in their original surroundings. A conventional curator produces interpretive materials in her office or library. In contrast, in Arvanitis' study the everyday curators composed their messages at the site of the monuments. Whereas the conventional curator's audience accesses the interpretive materials in the museum, the everyday curator's audience, Arvanitis, viewed the MMS messages through his own compatible mobile device. Writing instructors can appropriate the figure of the everyday curator for their own mobile composition assignments, thus allowing for alternative student discourses.

The second example of mobile composition centers on a trend called "sound-seeing." Sound-seers are people who record audio narrations of their travels or other sight-seeing activities—attending a grape festival, watching a ballgame, visiting a museum—and publish them on the Internet as podcasts. These podcasts may then be downloaded to a portable media player such as an iPod. The activity is called sound-seeing because the listener, or podcatcher, "sees" the event or place as it is described in "sound" by the podcaster. The term privileges the podcatcher's experience over that of the podcaster, whose activity may be dubbed "sight-sounding." The podcatcher can also hear the sounds associated with the object—the crushing of grapes, the roar of the crowd, museum visitors' comments—provided they are captured on the recording device. One sound-seeing web site invites audiences to "hear the sounds of wild streams and trees as [sic] I was in Belgium this past weekend" ("bicyclemark61_050823," 2005). Indeed, sound-seeing texts immerse the listener in the place of the speaker, whose intimate, material interactions with her surroundings are not just reflected in the writing but also, as in the example of mobile composition discussed above, constitute the subject of the writing. Whereas the independent distribution of "radio shows" over the Internet—to which podcasting technology was originally adapted—poses an alternative to the institutional discourses of radio and television, sound-seeing represents the development of podcasting into a new genre of mobile composition. This genre tends toward topics drawn from everyday life and aims to inform and entertain audiences. The space of writing—in this case, speaking—is the site of the object. Sometimes, sound-seers know in advance that they are going to visit, say, a farmer's market and record their impressions of it. At other times, they happen upon the market and decide to capture their impressions of it on the spot. In the former case, sound-seeing suggests a tourist's planned excursion. In the latter case, sound-seeing evokes the *flaneur*'s chance encounter with an object of interest. As with the case of the everyday curator, writing instructors can appropriate the figure of the tourist or *flaneur cum* guide for mobile composition assignments.

Recently, David Gilbert, a professor of communication at Marymount Manhattan College, and a group of his students created unofficial audio guides for the Museum of Modern Art and made them available as podcasts. Unlike the museum's official audio guide, which rent for $5, these guides are free and do not require the museum's audio wand, which is clunky by comparison with an iPod (Kennedy, 2005). The guides feature sardonic commentary from experts and nonexperts as well as soundtracks, inspired by the artworks, that sample symphonic themes, popular music, and 1950s television ads. Gilbert and his students call themselves Art Mobs,[7] and their website encourages visitors to send in their own audio guides for inclusion in the podcast feed. Whereas authors of sound-seeing texts compose them at the site of the object, Art Mobs designed their audio tour in conventional spaces, presumably campus computer labs. The opposite is true with regard to the space of reception. Sound-seers tend to listen to their podcasts in places other than those narrated in the text. In contrast, those who download Art Mobs' audio tour experience it onsite at the museum. Podcatchers can also enjoy Art Mobs' audio tour while viewing images of the artworks on the MoMA's website. Likewise, they can produce their own audio guides without ever visiting the MoMA. Dr. Gilbert and his students describe the guides on their website as a way to "hack the gallery experience" or "remix MoMA," while their larger goal is to "democratiz[e] the experience of touring an art museum" ("Art Mobs," 2005). Like Arvanitis' MMS image-texts, Art Mobs' podcast offers an alternative to the institutional discourse of the museum. In addition to the everyday curator, writing instructors can also appropriate the counter-curator as a figure for mobile composition assignments.

The third example of mobile composition focuses on the moblog, *mobile* and *blog*. A moblog consists of text, audio, photos, and video posted to a blog from a mobile device. The term is sometimes pronounced with the emphasis on the first syllable—MOBlog—out of affinity with the ideas about social self-organization developed in Howard Rheingold's *Smart Mobs* (2002). Rheingold predicted that advances in mobile technology would soon give everyone the tools to publish independent, real-time news reports directly on the Web and other platforms: "Imagine the power of the Rodney King video multiplied by the power of Napster. . . . Putting video cameras and high-speed Net connections in telephones, however, moves blogging into the streets. By the time this book is published, I'm confident that street bloggers will have constructed a worldwide culture" (pp. 168-169). The proliferation of moblogs shows that the culture of street bloggers that Rheingold predicted has indeed become a reality. Bryan Alexander (2004) cites an interesting academic example: "A team from

Umeå University in Sweden moblogged Jokkmokk's 399th Annual Sàmi Winter Market. Students applied their academic learning about the Sàmi to the real world, interviewing participants, conducting follow-up digital research on the fly, and uploading and expanding on commentary online" (p. 31).[8] However, the vast majority of moblogs focus on items of personal rather than public interest—photos of friends, the highway, the beach, and links to other blogs. Because inputting large amounts of text is a task better suited to keypads and styluses, what text, if any, moblogs contain is usually brief. Resembling an album of self-generated postcards, the typical moblog is a "record of travels in the world" (Hall, 2002, ¶ 27) characterized by quotidian details and seemingly superficial material interactions.

Thus, the question remains as to whether mobloggers will form "smart mobs," crowds of mobile technology users swarming to the same textual and physical locations to break important news in real time. (In the most famous example of a smart mob, a massive crowd of text-messaging protesters came together in a public square to bring down Philippine president Joseph Estrada in 2000 [Rheingold, 2002, pp. 157-158].[9]) Rheingold (2003) answers this question by stating that "the important remaining ingredient of a truly democratized electronic newsgathering is . . . a species of literacy—widespread knowledge of how to use these tools to produce news stories that are attention-getting, non-trivial, and credible" (¶ 18). Like all blogs, moblogs are both collaborative (audiences often double as authors) and accretive (the process is the product), but mobloggers are unique in that they publish at the location of the object (in Rheingold's citizen-reporter scenario, the news event). Journalistic moblogs aspire to provoke audiences into action as well as to pose an alternative to the institutional discourses of traditional news media. In imagining mobile phone users converging on the spot to cover important news while (and sometimes even before) it happens, Rheingold reconceives the street reporter and the mob, and writing instructors can appropriate both of these figures in their mobile composition pedagogies.

It is clear from these three examples that mobile composition assignments can take a variety of forms and involve a range of objects, places, and devices. Instructors might ask students to fashion themselves after one of the figures described or coalesce into intra-communicating groups in the service of a common social end. Such assignments reposition writers in the wild, where they must confront material conditions and respond to rhetorical opportunities not often encountered through traditional assignments. This new paradigm offers an alternative to the homework-fieldwork binary that dominates student writing today. In the homework model, students research and write in conventional spaces about objects located elsewhere.

In the fieldwork model, students research at the site of the object and write elsewhere. This binary separates the space of writing and publication from the object and tends to privilege work-in-progress genres such as essays and websites. On the other hand, mobile composition relocates writing and even publication in the place of the object and embraces process-as-product genres such as moblogs. We are not calling for the abandonment of all traditional writing experiences but for supplementing them with a new paradigm that draws attention to the materiality of writing. As unconventional writing spaces such as city parks, waiting rooms, and shopping malls grow familiar to students, conventional writing spaces such as the library, dorm room, and coffee shop will become defamiliarized, their very conventionality exposed. Instead of seeing a world of writing spaces and other spaces, students will see a world of tamed and untamed writing spaces. Ultimately, wherever they write, they will sense that they are "writing in the wild."

ENDNOTES

1. The past two decades have witnessed the emergence of a "new museology" (Peter Vergo, ed., *The New Museology* [London: Reaktion Books, 1989]) that privileges the avant-garde values Sirc considers vital to the future of composition studies. As a result, museums today have little in common with the museum discussed in Sirc's analysis, which is itself outdated insofar as it repeats many of the criticisms that since the mid-1990s have given way to more nuanced critiques of museum practice and theory.
2. Wi Fi (Wireless Fidelity) is a generic term for any of the wireless network protocols, such as IEEE 802.11b, that use unregulated radio or microwave spectrum.
3. There are two types of tablets: *slate* and *convertible*. Slates have touch-sensitive LCD screens with styluses and virtual keyboards, and convertibles add fully integrated keyboards. Separate keyboards can be attached to slate models.
4. These human factor issues may be remedied over time. In 2006, tablet weight ranges from 2-6 pounds, battery life from 2-6 hours, screens from 10-14 inches.
5. As mobile devices adopt 3G (third generation) specifications, they will also become location-aware. Students will be able to access maps, download information about their objects, and locate services such as ATMs, gas stations, and police stations.
6. MMS is a recent phenomenon in the United States, but smartphone users in Europe and Asia have been able to compose, send, and view messages that combine text, photos, audio, and/or video for several years now.
7. "Art Mobs" is an adaptation of "smart mobs," a concept discussed later in this section.

8. The URL for this moblog is http://blog.humlab.umu.se/jokkmokk2004/.
9. The recent outbreak of rioting in Paris demonstrated the destructive power of smart mobs. Many rioters coordinated their activities through blogs, cell phone calls, and texting.

REFERENCES

Alexander, Bryan. (2004). Going nomadic: Mobile learning in higher education. *Educause Review, 39*, 28-35. Retrieved February 12, 2008, from http://www.educause.edu/ir/library/pdf/ERM0451.pdf

Art Mobs to remix moma (with your help). (2005, May 11). Retrieved February 12, 2008, from http://mod.blogs.com/art_mobs/2005/05/art_mobs_to_rem.html

Arvanitis, Konstantinos. (2005). Museums outside walls: Mobile media and the museum in the everyday. In Pedro Isaias, Carmel Borg, Piet Kommers, & Philip Bonanno (Eds.), *Mobile learning 2005* (pp. 251-255). Proceedings of the IADIS International Conference, 28-30 June 2005, Qawra, Malta. Retrieved January 30, 2006, from http://www.le.ac.uk/museumstudies/ka43/kostas/publications. htm

bicyclemark61_050823. (2005, August 23). Bicyclemark's audiocommunique. Retrieved February 12, 2008, from http://bicyclemark.org/blog/2005/08/bicyclemark61_050823

Cohen, Jodi S. (2005, October 20). Missed class? Try a podcast. Digital recordings of lectures allow college students with MP3 players to catch lessons or just catch up wherever and whenever they want. *Chicago Tribune*. Retrieved February 12, 2008, from http://www.cs.duke.edu/news/?article = 167

Dean, Christopher, Hochman, Will, Hood, Carra, & McEachern, Robert. (2004). Fashioning the emperor's new clothes: Emerging pedagogy and practices of turning wireless laptops into classroom literacy stations. *Kairos, 9*(1). Retrieved February 12, 2008, from http://english.ttu.edu/kairos/9.1/binder2.html?coverweb/hochman_et_al/intro.html

Dobrin, Sidney I., & Weisser, Christian R. (2002). *Natural discourse: Toward ecocomposition*. Albany: SUNY Press.

Foster, Andrea L. (2003, April 4). Tablets sneak up on laptops. *Chronicle of Higher Education, 49*, A33. Retrieved February 12, 2008, from http://chronicle.com/free/v49/i30/30a03301.htm

Hall, Justin. (2002, November 21). From weblog to moblog. Retrieved February 12, 2008, from http://www.thefeaturearchives.com/24815.html

Hooper-Greenhill, Eilean. (1992). *Museums and the shaping of knowledge*. London; New York: Routledge.

Horner, Bruce. (2000). *Terms of work in composition: A materialist critique*. Albany: SUNY Press.

Kennedy, Randy. (2005, May 28). With irreverence and an iPod, recreating the museum tour. *New York Times*. Retrieved February 12, 2008, from http://www.nytimes.com/2005/05/28/arts/design/28podc.html?

Meeks, Melissa Graham. (2004). Wireless laptop classrooms: Sketching social and material spaces. *Kairos*, 9(1). Retrieved February 12, 2008, from http://english.ttu.edu/kairos/9.1/binder2.html?coverweb/meeks/index.html

Reynolds, Nedra. (2004). *Geographies of writing: Inhabiting places and encountering difference*. Carbondale: Southern Illinois University Press.

Rheingold, Howard. (2002). *Smart mobs: The next social revolution*. Cambridge, MA: Perseus.

Rheingold, Howard. (2003, July 9). Moblogs seen as a crystal ball for a new era in online journalism. *Online Journalism Review*. Retrieved February 12, 2008, from http://www.ojr.org/ojr/technology/1057780670.php

Sirc, Geoffrey. (2002). *English composition as a happening*. Logan: Utah State University Press.

Trimbur, John. (2001). *The call to write* (2nd ed.). New York: Longman.

Wysocki, Anne Frances. (2004). Opening new media to writing: Openings and justifications. In *Writing new media: Theory and applications for expanding the teaching of composition*. Logan: Utah State University Press.

Zoetewey, Meredith. (2004). Disrupting the computer lab(oratory): Names, metaphors, and the wireless writing classroom. *Kairos*, 9(1). Retrieved February 12, 2008, from http://english.ttu.edu/kairos/9.1/binder2.html?coverweb/zoe tewey/index.html

12

METAPHORS OF MOBILITY

EMERGING SPACES FOR RHETORICAL REFLECTION
AND COMMUNICATION

Nicole R. Brown

In the midst of a cultural code alien to you, what to do but transgress the code? In the midst of a city of signs that exclude you, what to do but inscribe signs of your own.

—Hal Foster, *Recoding: Art, Spectacle, Cultural Politics*

PLACE-BASED EDUCATION AND MOBILE TECHNOLOGIES

As rhetoric and composition teachers think critically about what it means for writing pedagogy to include mobility, there is a simultaneous focus in education on place-based learning and the grounding of curricula in local cultures, ecologies, geographies, and histories (Morgan, 2000; Sobel, 2004). The inclusion of *place* in pedagogy is well established in the field of rhetoric and composition, made visible through emerging concepts such as geographical rhetoric (Reynolds, 2004), ecocomposition (Weisser & Dobrin,

2001), and writing for/with the community (Deans, 2004). Simply stated, place-based writing aligns itself well with rhetoric's historical ties to citizens' shaping the fate of their communities (Halloran, 1993, p. 2). Furthermore, place-based pedagogies support the field's commitment to teaching critical literacy skills, which encourage students to reflect on local knowledge and to question "power relations, discourses, and identities in a world not yet finished" (Shor, 1997, p. 1). After all, as Peter McLaren and Henry Giroux (1990) affirm, a critical pedagogy "must be a pedagogy of place," tailored towards the specificities of the experiences, problems, languages, and histories that construct collective identities and possible transformations for communities (p. 163).

With the increased presence of wireless networks in educational, workplace, and social contexts, new understandings for what it means to write and distribute text in a place-based or context-aware manner are evolving daily. Currently, wireless connectivity enables users of mobile technology devices to connect to other users and the World Wide Web (WWW) in dynamic contexts. Whereas metaphors of mobility—such as the *nomad* or *frontier*—help to position our work and pedagogies in sites of broad impact, wireless networks when combined with location-aware technologies ask writing teachers and researchers to consider what our roles (and our students' roles) will be in defining both the contexts of composition *around* mobile technologies and the location-specific environments annotated by their use (Gillette, 2001). Through various forms of mobile, public reporting—museum and city audio tours, automobiles and hikers equipped with Global Positioning Systems (GPS) technologies, and even local radio broadcasting—these emerging rhetorical sites are already beginning to impact our teaching, research, and writing.

METAPHORS OF MOBILITY

To conceptualize these new, place-based yet mobile writing contexts, David Gillette (2001) argues for a critique of metaphors inherited from old composing spaces stating that "making this kind of virtual environment consistent with inherited metaphors such as windows, pages, and scrolls will probably not be possible" (p. 44). Most discussions on the use of metaphor in online contexts are structured around the visual language of graphical interface design like those identified by Gillette. Some of the most common or frequently encountered interface metaphors are that of the desktop, trash can, briefcase, or recycle bin, and as Cynthia L. Selfe and Richard J. Selfe (1994) explain, metaphors reveal political and ideological borders. Interface designers create such metaphors to assist users in making sense of their experiences with new software programs, gaming interfaces, or other unfamiliar environments or processes. Much different from the politically charged story told by Selfe and Selfe (1994), interface designers such as Alan Cooper (1995) view the employment of metaphor as a neutral strategy for making the interface coherent to users, therefore serving primarily cognitive purposes. Commenting on his own strategies regarding metaphor, Cooper (1995), however, distinguishes between discovering and applying metaphors for physical objects such as printers and documents and discovering and applying metaphors for "processes, relationships, services and transformations—the most frequent uses of software" (p. 61). For Cooper, metaphors related to processes, relationships, services, and transformations are more difficult to construct; it is in this category that metaphors for mobile composition lie.

Teachers of composition have started to think about, and stress the importance of, metaphors in emerging composing practices created by mobile technologies (Zoetewey, 2004). Although teachers of writing are certainly free to experiment with the naming of wireless sites for composition, the reality remains that we generally have little say about the metaphors defined by the developers of the computer-based programs. Selfe and Selfe's (1994) observation that "interface design will continue to be dominated primarily by computer scientists and will lack perspectives that could be contributed by humanist scholars" has indeed remained largely true (p. 498). However, as we [re]imagine writing assignments in technologically mediated contexts, we should pay close attention to relationships between the language used to describe and promote technologies and our own pedagogical practices and disciplinary values.

Although there are endless possibilities for articulating the pedagogical relevancy of students writing with mobile and location-aware technologies, a general statement might go something like this: The work of writers involves the critical, rhetorical, and social construction of culture/knowledge in contexts created, mediated, and circulated by "technologies" (including, but not limited to computing devices). Mobile and location-aware technologies offer a range of rhetorical situations around which we can conceptualize and create writing assignments. These assignments can invite students to construct place-based, public discourse; to foster rhetorical and critical inquiry; to write as a social act; and to view writing as a means to participate in new media literacies. As technologies become more mobile and location aware, Jay David Bolter and Richard Grusin

(2000) note that instead of placing the focus of our work on our computer screens, "the strategy of ubiquitous computer is to scatter computers and computational devices through our world" to "remediate space" (pp. 61-62). The metaphors of *graffiti* and *public art* may help compositionists build location-aware pedagogies. The metaphor of graffiti is applicable because it acknowledged the user's ability to annotate geographic space, providing information relevant to current geo-located contexts (Burrell & Gay, 2001), and the metaphor of public art provides for audience participation and rhetorical exploration of space. The metaphors of graffiti and public art, like most metaphors, provide teachers and researchers of writing with opportunities for both pedagogical participation and social critique and action.

GRAFFITI AS METAPHOR

Most contemporary understandings of graffiti have been established through a diverse range of graffiti subcultures that write inscriptions of identity (or graffiti) on local architectures and spaces for reasons ranging from political activism to no demonstrative purpose other than production and/or consumption (Scheepers, 2000, Introduction, ¶8). Unable to connect with greater art movements or forms of political activism, graffiti is not usually associated with meaningful or productive discourse. In general, graffiti writers inhabit space(s) between artist-citizen and vandal. In fact, for many graffiti writers, this contradictory space is essential to the work that graffiti writers do, which generally includes the placement of alternative discourse or transgressions on urban landscapes. The metaphor of graffiti, used to describe relationships between computer technologies and writing, first gained popularity through handheld Personal Digital Assistants (PDAs) such as the PALM Pilot. Because a PDA came without a keyboard, users applied universal strokes with a stylus to a touch-sensitive screen; these universal strokes became known as graffiti. More recently, graffiti has been used to describe relationships with technology extending beyond text styles. Researchers and developers on Cornell University's campus recently introduced a context-aware computing system named *Graffiti*, which uses the graffiti metaphor to describe its mobile services that enable users to "attach" electronic notes to specific locations on campus (Burrell & Gay, 2001, p. 1).

Graffiti is also being deployed to describe a mobile phone service in the United Kingdom called *TagandScan*, which enables subscribers to post "virtual graffiti" or "tags" to private grids in mobile phone zones. When sub-scribers log onto the service, they are able to view messages left by other subscribers in that zone and leave their own. Unlike the Cornell researchers—who took a wait-and-see approach to users employments of its mobile reporting services—the *TagandScan* website makes suggestions for place-based and public compositions that

- facilitate community development through reviews, histories, or general information;
- enhance spatial memory by creating maps, personal city guides, or shopping lists that travel along with you; and
- encourage public expression by creating opportunities for digital graffiti, political activism, or commemorating the events of a particular space. (Cimarrones Incorporated, 2003, Purpose section, ¶ 2)

The *TagandScan* website compares its service to gaining sixth sense, because current and former users' thoughts comingle with buildings and signs on public grids of information (Cimarrones Incorporated, 2003, ¶ 1). The suggestions for use, offered on the *TagandScan* website, expand the graffiti metaphor to include longer genres of geo-located writing. In ways markedly different from PALM Pilots', the developers of *TagandScan* employ the metaphor of graffiti to describe the act of posting information in and on specific objects and places. Considering the inherent social nature of this kind of writing, along with the ability for these genres to facil-itate critical inquiry and dynamic understandings of genre and literacy, the types of "tagging" identified by the promoters of *TagandScan*—community development, spatial memory, and public expression—could be used to construct writing assignments in wireless classroom or other mobile writ-ing contexts. In fact, it is likely that genres like those recommended on the *TagandScan* site—reviews, social histories, and activism—are already a part of many composition classrooms. Location-aware mobile technologies fur-ther encourage the teaching of these genres in ways that more directly relate them to space.

Linda Doyle, Fionnuala Conway, and Kenneth Greene (2003), a telecommunications research group in Dublin, Ireland, provide a range of composing possibilities that led them to use graffiti as metaphor to describe the digital writing context called *The Graffiti Wall*. The character-istics of the metaphor for *The Digital Graffiti Wall* that appealed to Doyle, Conway, and Greene include graffiti:

1. as a means of public protest (ranging from graffiti that simply defaces property to more elaborate organized political mural-like graffiti);
2. as a means of registering existence ("X was here" type statements);
3. as a means for engaging in dialogue (typically found on washroom doors through statements that get responses and answers); and
4. as a mean of opportunistic entertaining (to superimpose comments on existing public posters/images). (p. 2)

Although these defining characteristics establish rationales for writing that are similar to those proposed on the *TagandScan* web site, *The Digital Graffiti Wall* limits users to short [in length] compositions that rely not so much on words but more on the placement and the design of the text. Refashioning text to adapt to the inherently visual, rhetorical, and publishing contexts fits well with the work being done to consider the role of visuals in new or alternative writing genres (Kress & van Leeuwen, 1996; Wysocki, 1998; Yancey, 2004).

CRITIQUE OF GRAFFITI AS METAPHOR

Graffiti is useful for conceptualizing shared characteristics of place-based compositions, but there are sociocultural significances to the metaphor. Tony Silver and Henry Chalfant (1985), the writers and producers of *Style Wars*—a documentary of New York graffiti writers—identify the early subcultures of graffiti writers (particularly in New York City) as being intercultural and involving a range of classes. By comparison, the contemporary graffiti scene is populated by youths who live in lower socioeconomic conditions (not entirely, but mostly), making the city and urban decay a relevant canvas for these writers. As Tim Cresswell (1996) notes, graffiti writers are successful in their attempts to disrupt the cityscape by placing disorder in the landscape through transgressions that challenge existing orders of access (p. 38). Unlike most interface metaphors structured around "ideological axes that represent dominant tendencies in our culture" (Selfe & Selfe, 1994, p. 481), the metaphor of graffiti emerged from the margins— the very place where access to high-tech technology is least likely. Graffiti to describe place-based mobile composition can be problematic because of access issues—many taggers are not part of the high tech world.

Other limitations of the graffiti metaphor are made visible when the types or characteristics of writing encouraged by graffiti are considered alongside pedagogical goals related to writing instruction. For example, many of us have observed students constructing campus chalkings which function in similar ways to graffiti—they annotate space. Although campus chalkings sometimes represent pedagogical values related to writing as a social act, open access, public discourse, and genres as dynamic, many other campus chalkings lack critical and rhetorical perspectives. In similar ways to campus chalkings, graffiti writers "getting up" may eschew rhetorical perspectives.

In the case of graffiti being used to describe technologically mediated writing, it is important to remember that the metaphor acts both practically and persuasively, particularly through its relation to writing. Because graffiti is associated with edgy public discourse and social commentary, it has been co-opted to describe the writing spaces of location-aware mobile technologies. Although the metaphor describes some aspects of mobile and place-based writing, other forms of writing related to these technologies cannot be easily associated with graffiti. For example, graffiti does not necessarily invoke the collection and exchange of information connected to geographic and virtual spaces. David Gillette (2001) describes this writing as "useful content on devices that move through space, pick up information from the surroundings, react to the environment, metaphorically mimic the real world, and if we do our jobs well, help people navigate in both virtual and physical space" (p. 42). As Gillette's reflection on the changing nature of writing in the age of ubiquitous community reveals, writers face a new challenge to incorporate information onto, not just from, the world around them.

PUBLIC ART

In the past decade, the field of rhetoric and composition, and more specifically professional and technical communication, has well-articulated relationships between the purpose of a text and its design (Reddish, 2000; Sullivan, 2001; Wysocki, 1998). As Anne Frances Wysocki (2001) and Patricia A. Sullivan (2001) note, when it comes to elements of design, the field of rhetoric and composition would be foolish not to turn to the discipline of art for heuristics and principles. In the field of public art, the concept and act of graffiti is generally dismissed, despite the obvious craft of some works (Geer & Rowe, 1995, p. 24). According to Suvan Geer and

Sandra Rowe, the genre of graffiti seems to be plagued by association with "unsophisticated ubiquitous scrawls" (p. 24). Similar to graffiti subcultures, the field of public art places composition out of studios, off of canvases, away from computer screens or galleries and into our natural and built worlds; public art focuses on the interplays among form, audience, and placement.

The work of public artists is to promote art as inherently social-, place- and action-based. The field of public art works through critical understandings of relations among information, spaces, and location-aware mobile computing. According to Harriet Senie and Sally Webster (1992), public art involves a consideration of the "complex matrix in which it is conceived, commissioned, built, and, finally, received" (p. 3). Public art's focus on invention, publication space, the composing process, and reader response/participation fit well with pedagogical objectives related to writing instruction, especially writing as a social and critical act and the use and creation of dynamic, multiple genres and literacies.

Public art generally relates to instances when artist(s) and their subsequent work(s) of art establish one of two relationships with places and daily life: (a) they *integrate* with and become a part of the architecture or (b) they *intervene* with the architecture to raise social and political commentary. According to W.H.T. Mitchell (1990),

> A didactic emerges between . . . on the one hand art that attempts to raise up an ideal public sphere . . . (we might imagine here the classical image of a temple entrance or plaza filled with wise women and men engaging in enlightened discourse); on the other hand, art that disrupts the image of a pacified, utopian public sphere, that exposes contradictions and adopts an ironic, subversive relation to the public it addresses, and the public space where it appears. (p. 3)

Many cases of public art are, according to Malcolm Miles (1997), "somewhere between the polarities, partially aware of their political and social implications, yet adhering to conventional vocabularies of form" (p. 84). The Vietnam War Memorial is one such example (Mitchell, 1990, p. 3).

Similar to how public artists employ place-based sites for their compositions, the inclusion of location-aware mobile technologies in our writing pedagogies provide us with rhetorical contexts where students can thoughtfully consider relationships between themselves as writers and the architectures that construct their lives. *The Power of Place* (Hayden, 1995) in Los Angeles exemplifies how the concept of public art may be used to construct writing assignments around location-aware mobile technologies to encourage writing as a social and critical act. *The Power of Place* is a multidisciplinary project that draws on diverse and alternative histories and memories as a part of the conservation of local historic districts in Los Angeles—"to reclaim histories which have been obliterated by the dominant cultures" (Hayden, 1995, p. 177).

A similar location-based media project, *34 North, 118 West,* uses notebook computers and GPS technology to track users' locations through the Freight Depot in downtown Los Angeles. As participants move through sensor-tagged space, they participate in the co-construction of a story about space, time, architecture, and the industrial era (Knowlton, Spellman, & Hight, n.d.). Similar to the invention strategies provided by the developers of *TagandScan* or *The Graffiti Wall*, the range of compositions associated with *The Power of Place* and *34 North, 118 West* include types of composing already well-established by genres taught by writing teachers. The composition process, clearly established purposes and audiences, consideration of distribution, the high value placed upon audience participation, and the desire for both the re- and co-construction of space, all make public art a useful metaphor for composition studies.

Defining location-aware mobile computing technologies as either graffiti or public art means something is inevitably gained and something inevitably lost as we attempt to create composing practices using these metaphors. Focusing on the either-or also means we elide the broader question of why establish these connections at all. On the one hand, the open-access, grassroots capabilities of graffiti (or even public art or campus chalkings) have a certain appeal to those hoping for these new technologies to lead to social revolutions (Rheingold, 2004). For the purposes of gaining understandings of place-based writing—writing that is not just about a place but rather writing that is literally positioned as an integral part of the space to that place—both graffiti and public art are useful but in different ways.

The graffiti metaphor provides us with exigencies for writing. Graffiti is useful in articulating the various relationships that texts may have with specific geographic spaces and architectures, but the metaphor is situated in a sociocultural history that cannot and should not be dismissed. Despite graffiti writers' own commentary of their craft and its reliance on complex and related understandings of collaboration, language/image, intertextuality, as well as politicized understandings of urban space and dynamic genres, graffiti is dominantly associated with random or unsophisticated taggings and with criminal and destructive intent. However, graffiti when applied to mobile technologies explicitly reminds us of the links among socioeconomic status, political discourse, and issues of access. We can deploy this metaphor to help composition students critique issues of access, examine difference and power around the discourses and practices of mobile technologies, and even engage in the production of equitable mobile technology policy.

Public art is performative, always changing, generally well planned, and generated with the primary goal of composer and audience interaction within a particular space. Like graffiti, public art encourages critique and social action, but unlike graffiti, it does so by making the audience part of the performance. Public art also provides heuristics that call attention to how discourse—when combined with media platforms and architecture—can encourage action. Furthermore, the concept of public art fits well with values in the field of rhetoric and composition, because the production of public art is contextualized, inherently collaborative, process-based, and related to local rather than global narratives (Miles, 1997, p. 164).

Some characteristics of public art that may be used as invention strategies by composition teachers and students include: (a) writing as collaborative compositions involving multiple authors and intended audiences as authors; (b) writing focused on [re]defining space in ways that its [re]construction and use are illuminated; (c) writing as persuasive social action; (d) writing that makes visible how material space is capable of being read; (e) writing that involves the active exploration, critique, and investigation of space by writers and readers. Through emerging writing contexts brought forth by mobile computing and location-aware technologies, we can further advance questions regarding how student writers might, to use the words of one public artist, "create imaginative spaces in which to construct, or enable others to construct, diverse possible futures" (Miles, 1997, p. 84).

IN CONCLUSION

The creation and distribution of text is inherent in teaching composition, and location-aware mobile technologies ask us to rethink what it means to engage students in the (re)configuration of the world. When applied to describe technologically mediated communication, both graffiti and public art suggest ways to engage students in thinking about writing and its relationship to geographic spaces and discourses. Such composing calls attention the rhetorical context, creating exciting possibilities for positioning writing assignments as public discourse. These discourses may encourage the use of emerging and alternative genres and composing spaces in (and outside) our writing classrooms. Furthermore, as composition classrooms are extended into expanding contexts through the capabilities of wireless classrooms and expanding data clouds, we can ask students to situate themselves in geographic spaces of composition where they can conduct research and write in a particular time and within a particular space— reviews, directions, instructions, social histories, and so on. Put simply, the range of genres illuminated by place-based mobile technologies fits well with Kathleen Blake Yancey's (2004) call for the socially aware remediation of text to create new and dynamic genres and literacies.

REFERENCES

Acconci, Vito. (1990). Public space in a private time. In W.J.T. Mitchell (Ed.), *Art and the public sphere* (pp. 158-176). Chicago: University of Chicago Press.

Alexander, Bryan. (2003, April 7). Teaching in the wireless cloud. Retrieved December 20, 2004, from http://www.thefeature.com/article?articleid = 35265 &sh = alexander&ref = 5233460

Baron, Dennis. (1999). From pencils to pixels. In Gail E. Hawisher & Cynthia L. Selfe (Eds.), *Passions pedagogies and 21st century technologies* (pp. 15-33). Logan: Utah State University Press.

Batson, Trent. (1989). Teaching in networked classrooms. In Cynthia L. Selfe, Dawn Rodrigues, & William Oates (Eds.), *Computers, English and language arts: The challenge of teacher education* (pp. 247-256). Urbana, IL: National Council of Teachers of English.

Blakeslee, Ann M. (2001). Bridging the workplace and the academy: Teaching professional genres through classroom-workplace collaborations. *Technical Communications, 10,* 169-192.

Bloom, Lynn Z., Daiker, Donald A., & White, Edward M. (2003). *Composition studies in the new millennium: Rereading the past, rewriting the future.* Carbondale: Southern Illinois University Press.

Bolter, Jay David, & Grusin, Richard. (2000). *Remediation: Understanding new media.* Cambridge, MA: MIT Press.

Burrell, Jenna, & Gay, Geri K. (2001). *Collectively defining context in a mobile, networked computing environment.* Retrieved December 22, 2004, from http:// www.hci.cornell.edu/LabArticles/collective_burrell.pdf

Castleman, Craig. (1982). *Getting up: Subway graffiti in New York.* New York: MIT Press.

Cimarrones Incorporated. (2003). *TagandScan: Redefining reality.* Retrieved December 10, 2004, from http://www.tagandscan.com/

Cooper, Alan. (1995). *About face: The essentials of user interface design.* Foster City, CA: IDG Books Worldwide.

Cresswell, Tim. (1996). Imagining the nomad: Mobility and the postmodern primitive. In George Banko & Ulf Strohmayer (Eds.), *Space and social theory: Interpreting modernity and post modernity* (pp. 360-379). Oxford, UK: Blackwell.

Deans, Thomas. (2000). *Writing partnerships: Service learning in composition.* Urbana, IL: NCTE.

Deleuze, Gilles, & Guattari, Félix. (1986). *Nomadology: The war machine* (Brian Massumi, Trans.). New York: Semiotext(e).

Digiano, Chris, Yarnall, Louise, Patton, Charles., Roschelle, Jeremy, Tatar, Debra G., & Manley, Matt. (2003). Conceptual tools for planning for the wireless classroom. *Journal for Computer Assisted Learning, 19,* 284-297.

Doyle, Linda., Conway, Fionnuala, & Greene, Kenneth. (2003). Mobile graffiti, a new means of political engagement. In *Proceedings of the International Conference on Politics and Information Systems Technologies and Applications.* Orlando, FL: PISTA.

Foster, Hal. (1986). *Recoding: Art, spectacle, cultural politics.* Seattle: Bay Press.

Geer, Suvan, & Rowe, Sandra. (1995). Thoughts on graffiti as public art. *Public Art Review,* 24-26.

Gillette, David. (2001). Metaphorical confusion and spatial mapping in an age of ubiquitous computing. *Technical Communications, 48,* 42-47.

Green, Lelia. (2002). *Communication, technology and society.* London, UK: Sage.

Halloran, S. Michael. (1993). Afterthoughts on rhetoric and public discourse. In Victor Vitanza (Ed.), *Pre/text: The First Decade* (pp. 52-68). Pittsburgh: University of Pittsburgh Press.

Handa, Carolyn. (Ed.). (1990). *Computers and community: Teaching composition in the twenty-first century.* Portsmouth, NH: Boynton/Cook.

Hawisher, Gail E., & Selfe, Cynthia. (Eds.). (1999). *Passions pedagogies and 21st century technologies.* Logan: Utah State University Press.

Hayden, Dolores. (1995). *The power of place: Urban landscapes as public history.* Cambridge, MA: MIT Press.

Haynes, Cynthia, & Holmevik, Jan Rune. (2001). *High wired: On the design, use and theory of educational MOOs.* Ann Arbor: University of Michigan Press.

Hsi, Sherry. (2003). A study of user experiences mediated by nomadic web content in a museum. *Journal of Computer Assisted Learning, 19,* 308-319.

Jones, Steven G. (1998). Information, internet, and community: Notes toward an understanding of community in the information age. In Steven G. Jones (Ed.), *Cybersociety 2.0: Revisiting computer-mediated communication and community* (pp. 1-34). Thousand Oaks, CA: Sage.

Kent, Thomas. (Ed.). (1999). Introduction. In *Post-process theory: Beyond the writing process paradigm* (pp. 1-6). Carbondale: Southern Illinois University Press.

Knight, Will. (2003). Mobile phone service posts virtual graffiti. *New Scientist, 16.17.* Retrieved December 5, 2004, from http://www.newscientist.com/article.ns?id=dn4496

Knowlton, Jeff, Spellman, Naomi, & Hight, Jeremy. (n.d.). 34 north, 118 west. Retrieved December 27, 2004, from http://34n118w.net/

Kopelson, Karen. (2002). Dis/integrating the gay/queer binary: Reconstructed identity politics for a performative pedagogy. *College English, 65,* 17-35.

Kress, Gunther, & van Leeuwen, Theo. (1996). *Reading images: The grammar of visual design.* London, UK: Routledge.

Kress, Gunther, & van Leeuwen, Theo. (2001). *Multimodal discourse: The modes and media of contemporary communication.* New York: Oxford University Press.

Lo, Adrienne. (1998). *Native American students make voices heard.* Retrieved January 3, 2005, from http://www.yaleherald.com/archive/xxvi/10.16.98/news/voice.html

McLaren, Peter, & Giroux, Henry. (1990). Critical pedagogy and rural education: A challenge from Poland. *Peabody Journal of Education, 67,* 154-165.

Miles, Malcolm. (1997). *Art, space and the city: Public art and urban futures.* New York: Routledge.

Mitchell, W. J. T. (Ed.). (1980). Introduction. In *The language of images* (pp. 1-8). Chicago: University of Chicago Press.

Mitchell, W. J. T. (Ed.). (1990). Utopia and critique. In *Art and the public sphere* (pp. 1-5). Chicago: University of Chicago Press.

Moeller, Ryan. (2004). Wi-Fi rhetoric: Driving mobile technologies. *Kairos, 9*(1). Retrieved December 12, 2004, from http://english.ttu.edu/kairos/9.1/binder2.html?coverweb/moeller/index.html

Morgan, John. (2000). Critical pedagogy: The spaces that make a difference. *Pedagogy, Culture and Society, 8,* 273-289.

Nardi, A. Bonnie, & O'Day, Vicki L. (1999). *Information ecologies: Using technology with heart.* Cambridge, MA: MIT Press.

Penley, C., & Ross, A. (Eds). (1991). *Technoculture.* Minneapolis: University of Minnesota Press.

Redish, Janice C. (2000). What is information design? *Technical Communications, 47*(2), 163-166.

Reynolds, Nedra. (2004). *Geographies of writing: Inhabiting places and encountering difference.* Carbondale: Southern Illinois University Press.

Rheingold, Howard. (2002). *Smart mobs.* Cambridge, UK: Perseus Book Group.

Scheepers, Ilse. (2004). *Graffiti and urban space* (Honours thesis, University of Sydney, Australia, 2004). Retrieved December 15, 2004, from http://www.graffiti.org/faq/scheepers_graf_urban_space.html

Selber, Stuart. (2004). *Multiliteracies for digital age.* Carbondale: Southern Illinois University Press.

Selfe, Cynthia. L. & Selfe, Richard, J., Jr. (1994). The politics of the interface: Power and its exercise in electronic contact zones. *College Composition and Communication, 45,* 480-504.

Senie, Harriet F., & Webster, Sally. (1992). *Critical issues in public art: Content, context, and controversy.* New York: Harper Collins.

Shor, Ira. (1997). What is critical literacy? [Electronic version]. *The Journal of Pedagogy Pluralism & Practice, 4*(1). Retrieved January 10, 2005, from http://www.lesley.edu/journals/jppp/4/shor.html

Silver, Tony (Director), & Chalfant, Henry (Producer). (1985). Interview. In Tony Silver (Director), *Style wars*. RYKO/Palm Pictures.

Sobel, David. (2004). *Place-based education: Connecting classrooms & communities*. New York: Orion Books.

Soja, Edward W. (1989). *Postmodern geographies: The reassertion of space in critical social theory*. London, UK: Verso.

Soja, Edward W. (2001). In different spaces: Interpreting the spatial organization of societies. *Proceedings 3rd International Space Syntax Symposium Atlanta* [WWW document]. Retrieved from: http://undertow.arch.gatech.edu/homepages/3sss/papers_pdf/s1_Soja.pdf

Sullivan, Patricia A. (2001). Practicing safe visual rhetoric on the World Wide Web. *Computers and Composition, 18,* 103-121.

TagandScan. (2004). *TagandScan*. Retrieved December 18, 2004, from http://www.tagandscan.com/

Tuters, Marc. (2002). *Geograffiti locative GPSter: Theorizing the radical potential of location-aware mobiles*. Retrieved December 8, 2004, from http://www.gpster.net/potentialmobiles.html

Vielstimmig, Myka. (1999). Petals on a wet, black bough: Textuality, collaboration, and the new essay. In Gail E. Hawisher & Cynthia Selfe (Eds.), *Passions, pedagogies and 21st century technologies* (pp. 89-114). Logan: Utah State University Press.

Vincent, Norah. (2003). Campus chalkers make bogus free speech. *The Detroit News*. Retrieved January 3, 2005, from http://www.detnews.com/2003/editorial/0301/03/a09-49601.htm.

Weisser, Christian R., & Dobrin, Sidney I. (2001). *Ecomposition: Theoretical and pedagogical approaches*. Albany: SUNY Press.

Wysocki, Anne Frances. (1998). Monitoring order: Visual desire, the organization of web pages, and teaching the rules of design. *Kairos, 3*(2). Retrieved January 12, 2005, from http://english.ttu.edu/kairos/3.2/features/wysocki/mOrder0.html

Wysocki, Anne Frances. (2001). Impossibly distinct: On form/content and word/image in two pieces of computer-based interactive multimedia. *Computers and Composition, 18,* 137-162.

Wysocki, Anne Frances, & Johnson-Eilola, Johndan. (1999). Blinded by the letter: Why are we using literacy as a metaphor for everything else? In Gail E. Hawisher & Cynthia Selfe (Eds.), *Passions, pedagogies and 21st century technologies* (pp. 89-114). Logan: Utah State University Press.

Yancey, Kathleen Blake. (2004, December). Made not only in words: Composition in a new key. *College Composition and Communication, 56,* 297-328.

Zoetewey, Meredith. (2004). Disrupting the computers lab(oratory): Names, metaphors, & the wireless writing classroom. *Kairos, 9*(1). Retrieved December 12, 2004, from http://english.ttu.edu/kairos/9.1/binder2.html?coverweb/zoetewey/index.html

PART V

Teaching and Research in My Pocket: Mobile Gadgets and Portable Practices

13

THE GENIE'S OUT OF THE BOTTLE

LEVERAGING MOBILE AND WIRELESS TECHNOLOGIES IN QUALITATIVE RESEARCH

Clay Spinuzzi

Data Point 1: In May 2004, the *San Jose Mercury News* ran an article on how ubiquitous digital photography has furnished uncensored images of the conflict in Iraq, including the infamous Abu Ghraib photos. "You can't make all the cell phones go away," one officer in Iraq tells us. "You can't make all the digital cameras go away. The genie's out of the bottle" (quoted in Kewney, 2004a, ¶ 10).

Data Point 2: On the same day, the information technology website the *Register* reported: "It's always been something of an in joke with those who know the Japanese market for miniature cameras. You know that they are described as 'up-skirt' devices, but you always assumed this was a witticism. But no: and now it's going to get the camera phone made illegal in America. The Video Voyeurism Prevention Act has now been approved. It passed the Senate last year, and has now got through the House Judiciary Committee. It, or something like it, will almost certainly become law" (Kewney, 2004b, ¶ 2).

Data Point 3: Less than a month later, the *Register* reported that Sprint had produced a camera-less version of the TREO smartphone for enterprise use. Apparently businesses are nervous about their employees taking pictures of business secrets or other inappropriate things, so Sprint "blinded" the phones. As the *Register* dryly put it, "There's no word yet of Xerox producing a tonerless copier machine, which suggests that the fear of camera phones is more a matter of perception" (Orlowski, 2004, ¶ 3).

Data Point 4: At work, a child care provider pulls out her mobile phone to check her messages. One little boy brightens up and stands straight in front of her. "Cheese!" he says.

Data Point 5: The Pacific tsunami in December 2004, the London bombings in July 2005, and the flooding of New Orleans in August 2005 were all recorded by capture devices wielded by citizens: primarily camera phones and webcams. They were distributed through various means, including email, blogs, and news sites. In London in particular, officials turned from the dense network of government-controlled cameras and appealed to citizens to cull through their own digital photos for clues to the bombing. And in New Orleans, the citizen journalism (in text and pictures) contrasted positively to the appallingly inaccurate mainstream media reporting.

When you compile enough points, you start to see patterns. The pattern I'm seeing here is that information technologies, which have been shrinking, consolidating, and entangling at a breathtaking rate, are starting to have surprising impacts on the ways we collect, query, interpret, and validate information. My bottom-of-the-line mobile phone—the kind you get for free when you renew your contract—lets me check e-mail, surf the web, download ring tones, and take voice memos. I can text Google SMS to get phone numbers, driving directions, and other information, and I can also text anyone else with SMS (see Sun, 2004). If I had chipped in another 50 bucks, I could have gotten a model with these features, plus a built-in camera (which, as Data Point 4 suggests, has become a nearly ubiquitous feature). And if I had wanted to spend a few hundred dollars, I could have gotten a TREO—a PalmOS machine with more computing power than the PC on which I wrote my dissertation. That's a lot of computing power and functionality for someone to casually carry in his or her pocket. (Of course, it will sound quite trivial even by the time this chapter is published.)

This functionality is startling enough in itself, but what really makes it revolutionary is that it is networked. If camera phones just allowed us to take pictures of our children, they wouldn't be much different from Instamatics. But give any citizen the ability to instantly and cheaply publish those pictures where they can be accessed by anyone with an Internet connection, and you get some startling shifts. The results have led most immediately to a crisis in journalism, as "citizen journalists"—that is, people with mobile capture devices, a website, and an interest in telling stories—have destabilized that industry significantly. The genie is certainly out of the bottle, which is to say that information has circumvented the monopoly of news distribution through networked technologies. Those of us who are not journalists may be amused by, or even rooting for, bloggers who threaten with increasing frequency to "fact-check your ass."

You know who else might get fact-checked? Qualitative researchers.

EMPTY BOTTLES

Computing technologies have had a history of destabilizing industries. Initially that destabilization happened primarily in the form of automating skilled work and deskilling workers, nearly wiping out craftwork, further devaluing tacit knowledge, and thus serving Taylorist goals (Ehn, 1989; Zuboff, 1988). But computing technologies also refigured and distributed the means of production in various industries, and in so doing, undermined those industries by revealing how much of their specialization relied on their exclusive access to those tools. Graphic design, for instance, has never been the same since Broderbund's THE PRINT SHOP and other low-end design programs appeared. The results, of course, were often utterly horrendous: People with absolutely no design training took up simulated tools of graphic designers, producing documents with so many fonts that they looked like ransom notes. Eventually, those widely dispersed tools have helped to stabilize and transform graphic literacy across a larger population. And, not incidentally, they cheapened low-end graphic design work (which, given the tools, "anyone" could do) and differentiated it from high-end graphic design. Distributing the tools—the means of production—meant distributing (and arguably dissolving) much of the skills, expertise, and literacy involved in using them, and forced a significant transformation of the industry.

Similarly, journalism is reaching a crisis because its means of production (from printing presses to distribution networks to journalists' notebooks) are to some degree replicated by the combination of websites (particularly blogs), search engines, and mobile technologies. In the old model of journalism, journalists would gather information, editors would vet it according to stable disciplinary standards, and the press would print it. But that model has become destabilized because the two ends of the process turned out to be the anchors: Gatekeepers could control and enforce the standards primarily because they could control the means of production (Boczkowski, 2005, p. 164). These two constitute the bottle that holds the genie, or (to use another metaphor) a black box, Pandora's box (Latour,

1999). As long as that black box remains closed, we assume that its contents are stable and unitary. But the box can be forced open. As we found out in Rathergate,[1] once the means of production are distributed, anyone who has access to the new means of production can critique the old standards and introduce other standards from their own fields and disciplines. And the means of production have been so distributed that we are approaching the state that Michel Serres (1995) described in his discussion of the Greek agora:

> When everyone knows everything right now about everybody and lives by this knowledge, you have antiquity's notion of freedom and the ideal city, and also the ideal of modern philosophers since Rousseau, the ideal of the media and social science, of the police and bureaucracy: poll, clarify, inform, make known, expose, report. A terrifying nightmare, one that if you've lived in small villages or large tribes, you'll want to avoid all your life, for it is the height of enslavement. Freedom begins with the ignorance I have and wish to preserve of the activities and thoughts of my neighbor, and with the relative indifference that I hope they harbor for mine, for want of information. (p. 68)

That indifference, and the freedom that comes with it, is arguably coming to an end. When everyone has a digital camera, indifference gives way to pervasive, distributed surveillance. Notice that this is a very different model from the British one (see data point 5), in which the government's cameras view everyone and the government alone gets to pore through them to determine their significance. Less Panopticon, more agora.

As Serres argues, this distributed, pervasive model holds its own horrors. And in this case, they are most horrible to those who are in the position of the journalistic establishment: the reporters who take the photos and write the stories; the editors who vet them; the press operators who print them. These professionals have worked long and hard to keep Pandora's box closed. The distributed model destabilizes the industry because by distributing the means of production, it opens up the possibility of uncomfortable hybrids between disciplines, the mixing of disciplinary standards, the multiplicity of perspectives (Law, 2002; Mol, 2002) that the editorial vetting process is supposed to compress into a single perspective, a view from nowhere. The consumers also produce; those who have always been represented suddenly have the opportunity to represent themselves, to contest their representations, and to represent their interrogators. And the journalism industry, which has long seen itself as the "fourth estate," the spokesperson who represents those without a voice, finds that many of them suddenly do. As Michel Callon (1986) once said, "To speak for others is to first silence those in whose name we speak" (p.216). If no one will be silent, who can become the spokesperson, and how? If everyone holds the means of producing news, how can standards be stabilized to squeeze out multiplicity and produce a single story?

If that last question sounds a little far afield, let's apply it to qualitative research—a very different enterprise, but one with similarities that should make us nervous.

THE CORK WORKS FREE

For researchers, qualitative research has become an increasingly important tool for understanding the use of texts *in situ*, leading to an increasing focus on qualitative research in studies of classrooms and other educational programs, political action, and workplaces (e.g., Faber, 2002; Freedman, Adam, & Smart, 1994; Freedman & Smart, 1997; Spinuzzi, 2003a, 2003b; Winsor, 2001). Whether it is used informally (Henry, 1998; Wood & Silver, 1995), in structured design methods (Beyer & Holtzblatt, 1998; Millen, 2000; Viller & Sommerville, 2000), or in academic workplace studies (Doheny-Farina, 1986; Mirel, 1988; Spinuzzi, 2003b; Winsor, 2001), qualitative research has turned out to be useful for examining how people mobilize, interpret, and use texts.

Qualitative research can involve various techniques of data collection, but among the most frequent and useful are field notes, handwritten or typed notes in which the researcher records the participant's actions and the researcher's interpretations of them; interviews between the researcher and participant, typically audiotaped; and artifacts or sample objects that the participants used during the observation (Blomberg, Giacomi, Moser, & Swenton-Wall, 1993; Campbell, 1999; Creswell, 1997; Glesne, 1998; Wall & Mosher, 1994; for an engaging illustration of how these accumulate, see Latour, 1996). Researchers either collect surplus artifacts or photograph the actual artifacts that were used during the session.

Once these data are collected, the researcher files them away, then usually analyzes them later in several ways: describing, classifying, interpreting, representing, and visualizing (Creswell, 1997; see also Maxwell, 1996; Miles & Huberman, 1994; Strauss & Corbin, 1998). This analysis has traditionally involved laboriously cutting, shuffling, and coding actual paper, but it is increasingly done with qualitative research databases such as QSR N*VIVO, Qualis Research ETHNOGRAPH, and IdeaWorks QUAL-RUS. (For more on emergent coding in ethnographic databases, see Creswell, 1997; Fielding & Lee, 1998; Glesne, 1998; Miles & Huberman, 1994.) Because researchers must have access to the ethnographic database to perform the analysis, analysis is typically done in a defined workspace such as the researcher's office; analysis work can't be easily taken home for the weekend like other work, for instance. To facilitate their access to research materials, qualitative researchers have increasingly turned to mobile technologies such as laptops to store their data. The benefits of such technologies are so great that, after cautioning against using computer-based tools, some workplace researchers have come out with their own computer-based analysis tools (compare Beyer & Holtzblatt, 1998, with Holtzblatt, Wendell, & Wood, 2005).

In addition to data collection and analysis, qualitative research involves a considerable amount of record keeping related to project management. The qualitative researcher can expect to keep a calendar for scheduling observations, an address book for contacts, a set of notes for interview questions and other procedural issues, a list of administrative tasks to accomplish, and so forth (Strauss & Corbin, 1998). Traditionally, this record keeping is accomplished with a variety of paper-based genres. Such genres are adequate for single-researcher projects, but pose problems in terms of reproduction and sharing when it comes to research teams. Researchers are beginning to turn to mobile technologies, such as handheld devices, to digitally collect data that can be easily funneled into computer-aided qualitative data analysis systems (Blake, 2002; Pascoe, Ryan, & Morse, 2000; Spinuzzi, 2003c; Sun 2004).

That's the researchers' perspective.

To participants, however, qualitative research looks a lot like journalism. Someone comes by, records interviews, writes notes, takes pictures, collects artifacts, and goes away; later, they publish an account that may quote the interviewee and may not conform to what the interviewee recalls happening. The cycle is much longer, the readership is much narrower, and theoretically the analysis is more rigorous, but the configuration is essentially the same: qualitative data collection, followed by a process of analysis and vetting, followed by publication. Indeed, the data collection and publishing tools are often identical to those used in journalism! And although the analysis and vetting process is quite different in journalism and in qualitative research, it is protected, made stable and legitimate, in great part for the same reason: because it is protected by the two ends of the process, in which the researcher holds a monopoly. These ends constitute the bottle that constrains the genie.

That's not to say that research participants don't have a voice. Participants have been known to disagree sharply with the way they are represented in qualitative research. In one case, for instance, Ann Blakeslee and her coauthors (1996) described the negative response of a participant to an article Blakeslee published:

> He said that the argument I had made about the collaborative nature of his interactions with three scientists who read and responded to one of the physicists' drafts misrepresented what had occurred in that situation. He also said that my use of the term "collaboration" was too inclusive and, therefore, not useful. (p. 138)

In this case, Blakeslee responded by interviewing the participant for a subsequent article, and subsequently by meditating on how to draw participants into the process. The latter has become a recurring interest for qualitative researchers, who have increasingly positioned themselves as sympathetic listeners, participant-observers, who want to bring in the "voices" of those they interview. Those voices are brought in through a variety of methods: member checks, unedited responses to research (Winsor, 1996), subsequent response interviews (Blakeslee et al., 1996), or even co-authoring (Selfe & Hawisher, 2004). Other approaches have attempted to bring in participants in more direct ways. Participatory design, for instance, leverages several different techniques that encourage participants to directly transform the object of the research (Spinuzzi, 2005a, 2005b). But in these methods, the voices are invited and often stage-managed and edited. Like journalists, qualitative researchers must silence the participants in order to speak for them, to compress their multiplicity of perspectives into a view from nowhere. Like journalists, qualitative researchers truck in representations of their participants, representations that sometimes seem distorted to the participants themselves. And like journalists, qualitative researchers often see themselves as advocates for their participants, a sort of fourth estate that serves to rescue participants from their conditions (Spinuzzi, 2003b).

So the qualitative research process is largely designed and controlled by the researcher. And it is that design and control work—constituted on the front end by the initial research design, the conduct of the data collection, and the inscription of data in various research media, and on the back end by the publishing process—that keeps the enterprise coherent. We tend to think of the middle part of the process as the rigorous part: the analytical techniques, the summarization of implications, and the double-blind review.

But that part holds together only in the relatively protected environment created by the control over the means of production, which allow us to bracket other interpretations, analytical modes, and impressions. Like journalists, qualitative researchers often forget that without the relatively protected environment, this rigorous analytical center can be easily destabilized.

THE GENIE ESCAPES

Let's imagine by analogy what this might mean for the sort of qualitative research we often conduct in technical communication:

A researcher goes to a workplace to shadow and interview a participant. During the interview, the participant asks to record the conversation herself and takes pictures with a camera phone; she sees the visit as an interesting event that she wants to share with her friends. The researcher uneasily assents—it's fair, isn't it?

After completing the study, the researcher publishes an article in which he describes the workplace, its culture, and its foibles; this article in part includes some of the shadowing and interviews he conducted. A few days later, while ego-surfing (i.e., searching for his name using Internet search engines), he finds that the participant blogged before and after the observation—with photos and video (video? But he thought she had just been taking pictures!)— and her impressions are quite different. Worse, she found his article online, read it, and ridiculed it. The transcriptions that he had sweated over are compared to the audio she took and found to be inaccurate. His hesitations, uncertainties, and mistaken impressions, which he had excised from the published transcripts, are restored. Finally, the participant alleges that he had gotten key details wrong about the work and work organization; he has mixed up the official organization chart with the unofficial work organization, for instance.

And the damage can't be contained. When he Googles his name on Monday, the participant's site comes up as the fifth hit; by Friday, it's the first.

Has this nightmare scenario happened yet? Not to my knowledge. Will it? Oh yes. And when it does, I predict a reconfiguration of the researcher-participant relationship. Do we researchers enjoy seeing ourselves as participants' advocates now, as those who speak for the silent masses? What happens when we can't keep them silent, when they can "betray" us by representing themselves? Suddenly, involving participants in constructing research narratives is no longer a matter of noblesse oblige, but a necessary step in stabilizing research. Decentering the researcher indeed! It becomes vital to articulate a stance that takes participants into account, not as victims who need rescue or as people whose "voices" must be "heard," nor as antagonists, but as stakeholders who are entirely capable of holding a dialogue and even an argument. The qualitative researcher's story becomes just one story among others: a story that can be destabilized, just as researchers have long destabilized the stories of their participants by pointing out disjunctions among observations, interviews, and artifacts. All it takes is to let the genie out of the bottle—to distribute the means of data collection and publishing, the ends of the process. As Serres said: A nightmare.

Fortunately, some groundwork has already been laid for a response.

SPINNING THE BOTTLE

Research, John Law (2004) suggests, is conducted by setting up inscription devices for turning substances and events into traces and into chains of traces. Once these traces are generated, "the materiality of the process gets deleted" (p. 20). Our researcher, for instance, turns observations and interviews into notes and transcripts, then codes them with metadata that can themselves be retraced (categorized, quantified) to produce the findings at the heart of a research paper. This progressive chain of representations provides opportunities to see things that wouldn't be immediately obvious: patterns, redundancies, disruptions.

> Particular realities are being constructed by particular inscription devices and practices. Let me emphasise that: realities are being constructed. Not by people. But in the practices made possible by networks of elements that make up the inscription device—and the networks of elements within which the inscription device resides. The realities . . . simply don't exist without their matching inscription devices . . . such inscription devices—and even more so their particular products–are elaborate and networked arrangements that are more or less uncertain, more or less able to hold together, and more or less precarious. (p. 21)

Such inscriptions, and the practices they network, provide stable realities. Just as scientific work's "insecurity, typically invisible to outsiders, is apparent to anyone who visits a laboratory or knows anything about the actual conduct of science" (Law, 2004, p. 29), the work of qualitative research is messy and unstable until the work of inscription stabilizes it. That work must be black-boxed, made invisible, "deleted" for the stabilization to work; we have to have faith that the chain of inscriptions is sensible and competently done. If we have faith, subjectivity disappears (Law, 2004, p. 36). But different inscriptions, Law argues, result in different realities. There is a singularity, a single story or lesson to be derived, but it is an effect rather than a cause. Only when we delete the work can we see it as a cause (Law, 2004, p. 59). Is there a single reality? Yes—

> but only after the controversies have been resolved and the statements reporting on nature have become fixed, definite and unambiguous. Before this happens not only is reality indefinite, but at least at times of scientific controversy it is also multiple. Multiplicity is the product of the effect of different inscription devices and practices, for instance in different laboratories, producing different and conflicting statements about reality. Nevertheless, the end point—difficult but in their view none the less sometimes achieved in science—is a single reality and a single authorised set of inscription devices. (Law, 2004, p. 32)

As long as the researcher holds a monopoly on this chain of inscription devices, the research can largely be black-boxed and results can be reasonably certain. Subjectivity disappears and a singularity results, a single reality. But once someone else enters the lab, or investigates the chain of inscriptions, the black box flies open; the bottle is uncorked and the genie escapes. And then "different realities overlap and interfere with one another. Their relations, partially co-ordinated, are complex and messy" (Law, 2004, p. 61). Through this interference, researchers' stories lose their gloss, stability, and ethos.

But enough theory. Now we understand why the genie escaped. And perhaps we didn't know the difference between the genie and the bottle before, and like the journalists, we're not sure how to envision a world in which the genie has escaped. And all we have is an empty bottle, and all we can do is wonder how to recapture the genie: to reassert control, to silence the participants so we can speak for them. But there's an alternative, which is to leave the bottle empty. That is, we can come to terms with the fact that different inscriptions and different processes can result in different realities being enacted, and that our participants can and will produce their own inscriptions. If coordinated, this work can result in a "virtual singularity," a settlement that everyone can live with. But that settlement hinges on letting the other stakeholders speak for themselves.

This isn't an entirely new idea. For instance, Lucy Suchman (1995) has argued that participants should control their own representations. Participatory action research (PAR) has insisted on marrying research to the participants' concerns (Glesne, 1998). And PAR's stepchild, participatory design, was founded on the idea that participants should have an explicitly political voice in the research and reap substantial benefits from it (Ehn, 1989; Spinuzzi, 2005a, 2005b).

Participatory design in particular provides some interesting ways to encourage participants to become more involved in the research process. Those ways included organizational games (Bødker, Grønbaek, & Kyng, 1993, pp. 166-167), role-playing games (Iacucci, Kuutti, & Ranta, 2000), organizational toolkits (Bødker et al., 1987; Ehn & Sjögren, 1991; Tudor, Muller, & Dayton, 1993), future workshops (Bertelsen, 1996; Bødker, Grønbaek, & Kyng, 1993, p. 164), storyboarding (Madsen & Aiken, 1993), mockups (Ehn, 1989; Ehn & Sjögren, 1991; Bødker et al., 1987), and cooperative prototyping (Bødker & Grønbaek, 1991; Grønbaek & Mogensen, 1994): techniques that were designed to facilitate co-ownership of research with participants, to bring them into both data collection and analysis, and to invest them in the research. These techniques have turned out to be useful, but they have also tended to be organized and run by the researchers and have been too time-consuming to regularly deploy without the backing of other authorities (such as unions).

I suggest drawing inspiration from these techniques, but turning to the same technologies that helped to pop the cork in the first place. Mobile and networked technologies may have let the genie out of the bottle, but they also provide opportunities to invest participants in the research and let them coauthor the chain of transcriptions. By doing so—by bringing participants back into the research process as stakeholders, building on participatory research traditions—we can work toward a virtual singularity, not by shutting participants out and closing the process, but by opening up negotiations and creating new settlements.

GETTING OUR WISH

How do we distribute control over data rather than consolidating it, and how do we explore the multiplicity of perspectives rather than compressing them into a monologic (Panoptic) view from nowhere? Rather than per-forming analyses alone, in a closed system, it makes sense to perform them in an open system, a dialogic space in which many can participate. Less Panopticon, more agora.

As I discussed earlier, the impulse toward dialogic self-representation in research has taken many different forms. Some researchers have per-formed member checks (Maxwell, 1996; Winsor, 1996); others have employed techniques such as future workshops, organizational games, and prototyping (Greenbaum & Kyng, 1991); still others have enlisted partici-pants in collecting their own data (Hart-Davidson, 2003). But networked environments make this sort of data sharing and collaborative analysis simpler, more practical, and more communal. Just as blogs and camera phones allowed various actors to bring their experiences to bear on jour-nalism and to produce their own citizen journalism, an open system (Spinuzzi & Zachry, 2000) could allow participants to perform rolling mem-ber checks and to contribute to analyses in progress. Obviously the system couldn't be too open—analysis should be open to participants rather than the entire Internet community, and data should be managed carefully to conform with human subjects guidelines—but within those strictures, it makes sense to open the system more and invite participation in terms that participants find amenable, workable, and within their interests (see Spinuzzi, 2005a).

So what might that participation look like? It should facilitate participa-tory analysis in an open system, in which participants can examine each others' analyses, collaboratively reconstruct and examine their practices, and deliberate over future actions. And, as the data points suggest, this sys-tem can be—perhaps must be—heavily mediated through networked tech-nologies that situate the researcher's narrative as one among others. Let me suggest, half-facetiously, four open-system models.

Amazon.com: A Model for Member Checks?

At Amazon.com, readers are invited to rate and post comments on differ-ent products. This feedback helps other shoppers to make informed deci-sions based on peer comments rather than advertising, creating a wider dialogue about given products and providing the opportunity for decision making, comparisons, and strategy sharing among consumers.

Commenting on the products of others' work sounds a lot like mem-ber checks: points at which research participants read and comment on the drafts of qualitative researchers, establishing a dialogue in which they can represent themselves and provide their own interpretations of the researcher's work. Because the researcher's analysis is a narrative about her participants, the reasoning goes, those participants should have the chance to respond through their own counternarratives. (For an example, see Winsor, 1996.) In theory, member checks allow participants to achieve equal footing with the researcher; in practice, they tend to be framed by and overwhelmed by the researcher's voice. That's not what happens with Amazon product reviews, perhaps because those reviews encourage dialogue among customers. Could this model provide insight into how we might further dialogize member checks?

The model represents a shift from the Panopticon to the agora, emphasizing the community's self-monitoring and collective meaning making as participants make sense of the researcher's narrative, reducing it to the status of their own narratives. In doing so, participants could bring in their own perspectives and generate productive disagreements. As John Law (2004) suggests, bringing this "mess" back into research can encourage researchers—and participants—to generate and coordinate multiple perspectives of events without the expectation of being able to reconcile all of them.

Hotornot.Com: A Model for Inter-rater Reliability?

At hotornot.com, participants post their photographs and visitors rate those photos in terms of sexual attraction: are they hot or not? Visitors encounter a stream of different photos: once they rate a photo, they are told how their vote compared with others', and they are encouraged to rate the next photo. (Of course, some participants may have particular incentive to game the system. Who doesn't want to be considered hot?)

The rapid-fire rating system sounds a lot like a basic technique for measuring inter-rater reliability: it features a Likert scale, records responses, and compares those responses. At the same time, it opens up ratings to a larger population, and in a way that is less controlled. Could this model provide a practical, rapid way for participants to become raters of the researcher's analysis and to come to consensus about the data's meaning? Could it provide further avenues for dialogue that can both enhance the researcher's analysis and increase the participants' investment?

Again, this sort of measure could function in an expanded way to explore the multiplicity of perspectives held by participants. Where do participants agree and disagree? Do they tend to clump into groups, and if so, are those groups related to specific perspectives, social languages, or activities? Once the differences surface, can and will participants negotiate them?

Wikipedia.org: A Model for Participatory Analysis?

At Wikipedia, random participants come together to create a web-based encyclopedia. Anyone can create an entry. In fact, anyone can modify an existing entry, adding, correcting, or even falsifying information. The theory behind Wikipedia is that someone will always come along to correct mistakes and falsehoods; generally, the theory has held up quite well.

Applied to qualitative analysis, Wikipedia could support interesting variations on participatory data analysis. Like future workshops (Jungk & Mullert, 1988), organizational games (Bodker, Kensing, & Simonsen, 2004), and prototyping (Spinuzzi, 2005a), a Wikipedia-type model could provide a collaborative space in which participants and researchers participate on equal footing to analyze and write up research results. Could this approach inform or replace the practice that some qualitative researchers have followed, that of listing participants as coauthors (e.g., Selfe & Hawisher, 2004)? Could it provide an avenue for actually encouraging participants to generate and take ownership of analytical texts?

Friendster.com: A Model for Describing Participants and Their Relationships?

Friendster is a social networking application that allows users to describe themselves and their networks of associates. That is, it does work that we often find ourselves doing in qualitative research: describing participants and dimensions of their relationships with one another. But whereas qualitative researchers typically author the descriptions they create, Friendster allows participants to create and control their own representations. Groupings build organically and inductively.

Can a social networking application like Friendster allow participants to describe themselves and their groups in multiple ways that would be productive for research? Can participants use this sort of application to not only describe themselves but also examine and critique researchers' portrayals of their relationships?

CONCLUSION

In throwing out these half-serious (and in some cases half-baked) possibilities, I suggest ways in which qualitative researchers can come to grips with the increasing possibility that the genie will be let out of the bottle. We researchers are sometimes gripped with megalomaniac impulses: even when we're encouraging participation, we tend to stage-manage it so that our own analyses are strengthened rather than destabilized. This impulse to throw our arms around the data, control it, protect it, and wrestle some meaning out of it—that impulse is one we may no longer be able to indulge. In turning to these admittedly flawed models, I suggest that mobile and networked technologies don't necessarily have to lead to the same degree of individual control that is implied in your handheld device's PIM software. They can—and perhaps must—be used to set up dialogues and deliberations in which different participants with different perspectives can weigh in on the same phenomenon. We might even go further and encourage participants to research, document, and analyze the researchers themselves: watch the watchers, as citizen journalism has begun to do.

Obviously such models are not directly suitable for supporting collaborative analysis of qualitative data—you probably don't want participants rating each others' attractiveness—but they do provide us with some inkling of what a thoroughgoing open system might bring to participatory research. We can't put the genie back in the bottle, but maybe we can find a way to live with that.

ENDNOTES

1. "Rathergate" is the popular name for a scandal in which the news program 60 Minutes II broadcast a story about President Bush's service with the Texas Air National Guard. The story was partially based on documents that CBS reproduced on its website. Within hours, bloggers had raised a number of questions about the documents' authenticity based on typography, military terminology and conventions; in the next week, historical discrepancies and witnesses also cast doubt on the documents. CBS finally withdrew the claims, and an independent panel concluded that the report failed to meet CBS' standards of accuracy or fairness, though it stopped short of a definitive judgment about the documents' authenticity (Thornburgh & Boccardi, 2005).

REFERENCES

Bertelsen, Olav W. (1996). The festival checklist: Design as the transformation of artefacts. *Proceedings of the Participatory Design Conference,* Palo Alto, CA, pp. 93-101.

Beyer, Hugh, & Holtzblatt, Karen. (1998). *Contextual design: Defining customer-centered systems.* San Francisco: Morgan Kaufmann.

Bjerknes, Gro, Ehn, Pelle, & Kyng, Morten. (Eds.). (1987). *Computers and democracy—A Scandinavian challenge.* Aldershot, England: Avebury.

Blake, Edwin H. (2002). A field computer for animal trackers. *Proceedings of the ACM CHI 02 Conference on Human Factors in Computing Systems,* pp. 532-533.

Blakeslee, Ann M., Cole, Caroline M., & Conefrey, Theresa. (1996). Evaluating qualitative inquiry in technical and scientific communication: Toward a practical and dialogic validity. *Technical Communication Quarterly, 5,* 125-149.

Blomberg, Jeanette, Giacomi, Jean, Mosher, Andrea, & Swenton-Wall, Pat. (1993). Ethnographic field methods and their relation to design. In Douglas Schuler & Aki Namioka (Eds.), *Participatory design: Principles and practices* (pp. 123-156). Hillsdale, NJ: Erlbaum.

Boczkowski, Pablo J. (2005). *Digitizing the news: Innovation in online newspapers.* Cambridge, MA: MIT Press.

Bodker, Keld, Kensing, Finn, & Simonsen, Jesper. (2004). *Participatory IT design: Designing for business and workplace realities.* Boston: MIT Press.

Bødker, Susanne, Ehn, Pelle, Kammersgaard, J., Kyng, Morten, & Sundblad, Yngve. (1987). A UTOPIAn experience: On design of powerful computer-based tools for skilled graphical workers. In Gro Bjerknes, Pelle Ehn, & Morten Kyng (Eds.), *Computers and democracy—A Scandinavian challenge* (pp. 251-278). Aldershot, England: Avebury.

Bødker, Susanne, & Grønbaek, Kaj. (1991). Cooperative prototyping: Users and designers in mutual activity. *International Journal of Man-Machine Studies, 34*(3), 453-478.

Bødker, Susanne, Grønbaek, Kaj, & Kyng, Morten. (1993). Cooperative design: Techniques and experiences from the Scandinavian scene. In Douglas Schuler & Aki Namioka (Eds.), *Participatory design: Principles and practices* (pp.157-176). Hillsdale, NJ: Erlbaum.

Callon, Michel. (1986). Some elements of a sociology of translation: Domestication of the scallops and the fishermen of Saint Brieuc Bay. In John Law (Ed.), *Power, action and belief: A new sociology of knowledge?* (pp. 196-233). Boston: Routledge.

Campbell, Kim S. (1999). Collecting information: Qualitative research methods for solving workplace problems. *Technical Communication, 46,* 532-545.

Creswell, John W. (1997). *Qualitative inquiry and research design: Choosing among five traditions.* Thousand Oaks, CA: Sage.

Doheny-Farina, Stephen. (1986). Writing in an emerging organization: An ethno-graphic study. *Written Communication, 3,* 158-185.

Ehn, Pelle. (1989). *Work-oriented design of computer artifacts.* Hillsdale, NJ: Erlbaum.

Ehn, Pelle, & Sjögren, Dan. (1991). From system descriptions to scripts for action. In Joan Greenbaum & Morten Kyng (Eds.), *Design at work: Cooperative design of computer systems* (pp. 241-268). Hillsdale, NJ: Erlbaum.

Faber, Brent D. (2002). *Community action and organizational change.* Carbondale: Southern Illinois University Press.

Fielding, Nigel G., & Lee, Raymond M. (1998). *Computer analysis and qualitative research.* Thousand Oaks, CA: Sage.

Freedman, Aviva, & Smart, Graham. (1997). Navigating the current of economic policy: Written genres and the distribution of cognitive work at a financial insti-tution. *Mind, Culture, and Activity, 4,* 238-255.

Freedman, Aviva, Adam, Christine, & Smart, Graham. (1994). Wearing suits to class: Simulating genres and simulations as genre. *Written Communication, 11,* 193-226.

Glesne, Corrine. (1998). *Becoming qualitative researchers: An introduction* (2nd ed.). New York: Allyn & Bacon.

Greenbaum, Joan, & Kyng, Morten. (Eds.). (1991). *Design at work: Cooperative design of computer systems.* Hillsdale, NJ: Erlbaum.

Grønbaek, Kaj, & Mogensen, Preben. (1994). Specific cooperative analysis and design of general hypermedia development. In Randall Trigg, Susan Irwin Anderson, & Elizabeth Dykstra-Erickson (Eds.), *Proceedings of the Participatory Design Conference,* Palo Alto, CA, pp. 159-171.

Hart-Davidson, William. (2003). Seeing the project: Mapping patterns of intra-team communication events. *Proceedings of the ACM SIGDOC 2003 Conference,* pp. 28-34.

Henry, Pradeep. (1998). *User-centered information design for improved software usability.* Boston: Artech House.

Holtzblatt, Karen, Wendell, Jessamyn Burns, & Wood, Shelley. (2005). *Rapid contex-tual design: A how-to guide to key techniques for user-centered design.* San Francisco: Morgan Kaufmann.

Iacucci, Giulio, Kuutti, Kari, & Ranta, Mervi. (2000). On the move with a magic thing: Role playing in concept design of mobile services and devices. *Proceedings of the ACM DIS '00,* pp. 193-202.

Jungk, Robert, & Mullert, Norbert. (1988). *Future workshops: How to create desirable futures.* London, England: Institute for Social Inventions.

Kewney, Guy. (2004a). Did Rumsfeld ban Iraq camera phones? Retrieved April, 2006, from http://www.theregister.co.uk/2004/05/25/iraq-camera_phone_ban/

Kewney, Guy. (2004b). Up-skirt law to destroy mobile phone biz? Retrieved April, 2006, from http://www.theregister.co.uk/2004/05/17/up_skirt_law/

Latour, Bruno. (1996). *Aramis, or the love of technology.* Cambridge, MA: Harvard University Press.

Latour, Bruno. (1999). *Pandora's hope: Essays on the reality of science studies.* Cambridge, MA: Harvard University Press.

Law, John. (2002). *Aircraft stories: Decentering the object in technoscience.* Durham, NC: Duke University Press.

Law, John. (2004). *After method: Mess in social science research.* New York: Routledge.

Madsen, Kim Halsikov, & Aiken, Peter H. (1993). Experiences using cooperative interactive storyboard prototyping. *Communications of the ACM, 36*(4), 57-66.

Maxwell, Joseph A. (1996). *Qualitative research design: An interactive approach.* Thousand Oaks, CA: Sage.

Miles, Matthew B., & Huberman, A. Miles. (1994). *Qualitative data analysis: An expanded sourcebook* (2nd ed.). Thousand Oaks, CA: Sage.

Millen, David R. (2000). Rapid ethnography: Time deepening strategies for HCI field research. *Proceedings of the ACM DIS '00*, pp. 280-286.

Mirel, Barbara. (1988). The politics of usability: The organizational functions of an in-house manual. In Stephen Doheny-Farina (Ed.), *Effective documentation: What we have learned from research* (pp. 277-298). Cambridge, MA: MIT Press.

Mol, Annemarie. (2002). *The body multiple: Ontology in medical practice.* Durham, NC: Duke University Press.

Orlowski, Andrew. (2004). PalmOne blinds Treo smartphone for Sprint. Retrieved April 2006, from http://www.theregister.co.uk/2004/06/09/treo_no_camera/.

Pascoe, Jason, Ryan, Nick, & Morse, David. (2000). Using while moving: HCI issues in fieldwork environments. *ACM Transactions on Computer-Human Interaction, 7*(3), 417-437.

Selfe, Cynthia L., & Hawisher, Gail E. (2004). *Literate lives in the information age: Narratives of literacy from the United States.* Mahwah, NJ: Erlbaum.

Serres, Michel. (1995). *The natural contract.* Ann Arbor: University of Michigan Press.

Spinuzzi, Clay. (2003a). Knowledge circulation in a telecommunications company: A preliminary survey. *Proceedings of the ACM International Conference on Documentation*, pp. 178-183.

Spinuzzi, Clay. (2003b). *Tracing genres through organizations: A sociocultural approach to information design.* Cambridge, MA: MIT Press.

Spinuzzi, Clay. (2003c). Using a handheld PC to collect and analyze observational data. *Proceedings of the ACM International Conference on Documentation*, pp. 73-79.

Spinuzzi, Clay. (2005a). Lost in the translation: Shifting claims in the migration of a research technique. *Technical Communication Quarterly, 14*, 411-446.

Spinuzzi, Clay. (2005b). The methodology of participatory design. *Technical Communication, 52,*163-174.

Spinuzzi, Clay, & Zachry, Mark. (2000). Genre ecologies: An open-system approach to understanding and constructing documentation. *ACM J. Comput. Doc., 24*(3), 169-181.

Strauss, Anselm, & Corbin, Juliet M. (1998). *Basics of qualitative research: Techniques and procedures for developing grounded theory* (2nd ed.). Thousand Oaks, CA: Sage.

Suchman, Lucille A. (1995). Making work visible. *Communications of the ACM, 38*(9), 56-64.

Sun, Huatong. (2004). *Expanding the scope of localization: A cultural usability perspective on mobile text messaging use in American and Chinese contexts.* Unpublished dissertation, Rensselaer Polytechnic University..

Thornburgh, Dick, & Boccardi, Louis D. (2005). *Report of the independent review panel.* Retrieved April, 2006, from http://wwwimage.cbsnews.com/htdocs/pdf/complete_report/CBS_Report.pdf.

Tudor, Leslie, Muller, Michael, & Dayton, J. Thomas. (1993). A C.A.R.D. game for participatory task analysis and redesign: Macroscopic complement to PICTIVE. *INTERCHI '93*, pp. 51-52.

Viller, Stephen, & Sommerville, Ian. (2000). Ethnographically informed analysis for software engineers. *International Journal of Human-Computer Studies, 53,* 169-196.

Wall, Patricia, & Mosher, Andrea. (1994). Representations of work: Bringing designers and users together. In Randall Trigg, Susan Irwin Anderson, & Elizabeth Dykstra-Erickson (Eds.), *Proceedings of the Participatory Design Conference*, Palo Alto, CA, 87-98.

Winsor, Dorothy A. (1996). *Writing like an engineer: A rhetorical education.* Mahwah, NJ: Erlbaum.

Winsor, Dorothy A. (2001). Learning to do knowledge work in systems of distributed cognition. *Journal of Business and Technical Communication, 15,* 5-28.

Wood, Jane, & Silver, Denise. (1995). *Joint application development.* New York: John Wiley & Sons.

Zuboff, Shoshanna. (1988). *In the age of the smart machine: The future of work and power.* New York: Basic Books.

14

WINGED WORDS

ON THE THEORY AND USE OF INTERNET RADIO

Dene Grigar

John Barber

Throughout the *Odyssey*, Odysseus' "winged words," called *epea pteroenta* in ancient Greek, sustain him in his journey and gain him great gifts from gods and men alike. Although this epic has come to represent what is left of an ancient, lost culture, the notion of well-crafted and passionate words, spoken aloud and intended to be heard by a listening audience, still remains. One iteration of winged words made possible by broadband networks is internet radio.

Imagine this scenario: You have created a digital music composition. You want to perform it live to a group of people in New York City. To do so, you pack up your equipment, fly to the United States, and find a space to perform, and so on. . . .Or, you send your friends an e-mail message notifying them when you plan to perform and the URL for the site where they can listen in. They go to the site and listen to you perform live on the radio or download the file and play it later on their Apple iPod or cellular phone. In either case, words take flight; their mobility, assured.

With this idea in mind, our essay describes the Nouspace Internet Radio project (see Figure 14.1). This project entails using internet radio to enhance undergraduate- and graduate-level rhetoric and literature courses by broadcasting poetry and prose, music and interviews, and simultaneous Web showings of electronic literature and other interactive works. Although there are many internet radio programs that provide a wide range of infor-

Figure 14.1. Nouspace Internet Radio Project

mation, music, and news, few focus on the artistic output of digital or new media artists or intellectual discussions and responses to such artifacts. Sponsored by MoMA and P.S.1, the 24-hour station WPS1 is a notable exception ("Radio Art: Is there Life After Death?"). It stands to reason, therefore, that our Nouspace Internet Radio project has been of interest to traditional poets, fiction writers, and artists interested in moving their creative expressions into the larger realm of digital media, thereby gaining a wider audience for their work. But our project can also be useful for writing and rhetoric teachers eager to experiment with new ways to collaborate and communicate at a distance. Thus, in this essay we provide background on electronic technology relating to sound and aurality, outline our project and detail the theories underlying it, and talk about how writing teachers working in face-to-face and online scenarios can adapt this technology to their own classrooms.

BACKGROUND ON ELECTRONIC TECHNOLOGY, SOUND, AND AURALITY

Futurist Isaac Asimov (1991) argues that four fundamental advances in human communication have evolved and permanently altered every facet of our world: speaking, writing, printing, and computing. Each, he says, has worked to bridge the feelings of distance between peoples and to facilitate the notion of communication between communities and social segments (Asimov, 1991, pp. 426-431). However, over the last decade, based on the popularity of "computing"-based opportunities, numerous digital and new media writers and artists have rushed to colonize the electronic spaces of the Internet and the World Wide Web (WWW). This movement has resulted in a growing interest in aurality, evident in the presence of internet radio, the growing popularity of MP3 players—such as the iPod—and podcasting, and homemade music created with GUI-based programs such as Garage Band. Although we acknowledge that the move online has brought about opportunities to morph text and images, to create moving symbols for communication, to develop new ways of presenting or performing in an interactive context, to construct the reader-viewer as an active participant, and to engage in the art of music making and sound production, we also recognize that what has been forgotten, or at least left aside, are the sound and power of the author's voice, her rhetorical skills to inform, persuade, or convince, the aural quality of her dance with language and ideas, and the immersive quality of sound and its ability to facilitate interaction with an audience. What is often missing is *speaking*, the winged words.

We note that many of the current web-based and electronic iterations of poetry and prose appearing online are absent the voice of their authors, despite the fact that much of this work may have been created with the intent that it be heard by an audience and that much meaning and intent may have been designed to be communicated by the author's voice, her cadence, tone, emphasis, and enunciation. These works without voice are even despite the fact that, according to scientist Sidney Perkowitz (2004), "hearing [is] . . . our second most important sense, after vision" (p. 148). At one time the sound of an author's (the speaker) voice was highly valued. The rhetors of ancient Greece, striding into the *agora*, hands raised in the universal signal of their intent to speak, commanded respect and often swayed nation states with the power of their voices. Since then, leaders, warriors, and explorers have inspired their followers with the skillful use of their voices. Now, today, our words are mediated (see Figure 14.2) by elec-

Figure 14.2. Mediated Words

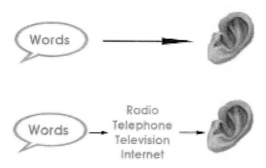

tronic technologies such as radio, telephone, television, and the broad-band frontier of global networks as seen in chat programs with voice potential like iChat and even voice service for broadband (VoIP). Yet even filtered through such technologies, the power of the speaker's voice to immerse us listeners in the events and activities of a narrative remains. In fact, author Ray Bradbury (1992) has called radio "image-making for the eardrums" (p. 4), and the long list of extraordinary radio broadcasts that created a sense of immersion in different realities for their audiences bears out his statement.

Witness the account by Herb Morrison, a reporter for Chicago radio station WLS, of the May 6, 1936, trans-Atlantic arrival of the German dirigible *Hindenburg* at Lakehurst, New Jersey. The rigid airship approached its mooring mast and exploded into a giant fireball. As the tragedy unfolded, in less than 30 seconds, Morrison's smooth delivery lapsed into a jumbled series of impressions, and his reporter's objectivity gave way to subjective horror. The result was that those listening achieved a sense of immediacy and nearness to an otherwise distant event—and Morrison lost his job.

Another excellent example was the October 30, 1938, radio dramatization of H.G. Wells' novella *The War of the Worlds* in which the voices and sound effects of Orson Welles and other players in his *Mercury Theatre on the Air* produced startling and unforeseen results: namely, that at least 1.2 million listeners took the broadcast literally and believed Earth to be undergoing invasion by beings from the planet Mars (Cantril, 1940). These alien beings were described in "winged words" in this way: "minds that are to our minds as ours are to the beasts in the jungle, intellects vast, cool, and unsympathetic, regarded this earth with envious eyes and slowly and surely drew their plans against us" (Koch, 1971, p. 36).

The adaptation of Wells' novel for radio broadcast reflected what was happening in Europe at the time. Soon afterward World War II began, and the sense of urgency and *gravitas* in the voices of Adolf Hitler and Winston Churchill, respectively, convinced masses of people to believe, to support, and to act. Dorothy Thompson, one of the few female radio reporters at the time, detailed negotiations between Great Britain and Germany regarding the German invasion of Poland on September 3, 1939. On American shores, the voice of President Franklin D. Roosevelt, heard via radio in homes across the nation, moved people from inertia and despair to confident action as he described the surprise December 7, 1941, Japanese attack on Pearl Harbor and sought a declaration of war.

Television, a visual medium, also has had its moments with sound: Dwight D. Eisenhower speaking about the dangers of the "military-industrial complex"; Billie Jean King taunting a chauvinistic Bobby Riggs; John F.

Kennedy declaring his solidarity with Berliners; Gloria Steinem enjoining women in the battle for their liberation; Martin Luther King exclaiming to the nation the famous words: "I have a dream"; Joan Baez singing to crowds of antiwar protestors; and Neil Armstrong emotionally claiming to the world, "One small step for man; one giant leap for mankind." All these examples speak to what we mean by Homer's "winged words." Communication technologies have extended their broadcast range, making spoken words available to audiences far removed by both time and space. But, as we have said, the voice of the speaker is curiously absent in many creative literary arts endeavors within the interactive spaces fostered by networked computer technologies, and radio art has become what *Net Art News* calls "a neglected medium" ("Radio Art: Is There Life After Death?").

NOUSPACE INTERNET RADIO AND THEORIES OF SOUND AND AURALITY

Let us now provide some background information about our project: We founded Nouspace Internet Radio in Fall 2003 in order to disseminate information for the purpose of scholarship and artistic expression to an at-a-distance audience in an aural context. It is part of a larger Nouspace Media Arts project that began with a MOO that Dene founded, called Nouspace, in Fall 2001 and includes web work, net art, and multimedia experiences.

John set up the station on our own server—an ordinary Mac G4—using a software program called Nicecast (see Figure 14.3). The program was inexpensive, about $40 U.S., easy to download, install, and configure. ROGUE AMOEBA Nicecast allows us to broadcast "anything you can plug into your Mac can also be broadcast out," including Apple iTunes or your own audio files (Rogue Amoeba, ¶ 4). The sound quality of our broadcast is scalable from monaural spoken voice to stereophonic CD music quality. Additionally, we can overlay live voice and various sound effects. Given the capabilities of this program, we were able to start broadcasting immediately, without the necessity of purchasing any additional or specialized equipment. In dividing up the responsibilities for the station, John's training and experience in radio and TV broadcast has landed him production responsibilities, and Dene handles design, program development, and grant writing.

To house the project, Dene built a web site that provides technical information, background on the project, and archives of events. Visitors to the site can download recorded MP3 files taken from the various events

Figure 14.3. Nicecast

and listen to them streamed from the Web or download them to listen to later on their MP3 player or cell phone—a phenomenon known as podcasting (Affleck, 2006, p. 5). Others in this volume, such as Olin Bjork and John Pedro Schwartz and Lisa Meloncon and Beth Martin, discuss podcasting and wireless technologies in greater detail. What should be obvious is that the mobility offered by such a technology not only extends the broadcast capability of internet radio but also the life of the word itself.

The bottom line is that with minimum investment anyone can operate a radio station and broadcast and/or podcast its content. These technologies also hold great pedagogical potential. For example, Dene has used Nouspace Internet Radio to teach orality and oral cultures in her graduate media and literature course. Drawing from stories archived at the National Museum in Sydney, Australia, she read Aboriginal myths as the content of one broadcast. Her students tuned in, listened to, and experienced these Aboriginal works as they had originally been presented: as spoken stories preserved through their vocal retelling. The experience helped graduate students understand the differences between oral and print cultures, such as the lack of kinesthetic involvement required of a "listener" as compared to a "reader." In addition, because some students lost their internet connection during the event, we were able to talk about transmission and communication models. In particular, the lost connection taught a valuable lesson on Claude Shannon's concept of "noise," or the "unexplained variation in a communication channel or random error in the transmission of information" ("Shannon-Weaver Model") to its intended audience (see Figure 14.4; for critiques of this model see also http://www.cultsock.ndirect.co.uk/MUHome/cshtml/introductory/trancrit.html).

Theoretically speaking, internet radio is representative of new media in that it combines digital technologies of networked culture with oral practices associated with live analog broadcasts and the spoken word. Internet radio suggests a transformation in the relationship between the user and information, a transformation that can be described as an epistemological shift requiring knowledge gained from not only the eye but also the ear or what British theorist Alan Beck calls "listening-in and knowing and how listening might be experienced as more 'real' than seeing" (Beck, 1999, Section 1.2, ¶ 2). This combination of seeing and listening lends itself well to explorations of artistic and scholarly practices. Although it is true that common communication technologies, such as radio and telephone, have retained, in part, aural qualities—Web companion pages and text messaging, notwithstanding—contemporary culture is generally described as a visual one. The academic world, in particular, is still solidly grounded in the paradigm of print. Thus, the value of the Nouspace Internet Radio project is that it provides a location for theorizing nonvisual models of representation for both artistic expression and scholarly modes of communication.

Imagine students being able to listen to a poet read her own work and then talk about the inspiration behind it, while they simultaneously read the text of that work on a web site—as well as accessing critiques, biographical information, and bibliographical resources about the author and her writing. Imagine accessing this material within a collaborative, interactive environment such as a MOO, or extended through downloadable MP3 files for future listening and study on players and cell phones. Imagine the new dimensions and power that new media might explore and use when such

Figure 14.4. The Shannon-Weaver Model

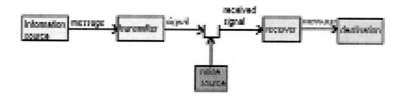

materials are available for listening from anywhere within students' wireless network or service provider territory. In essence, these are the kinds of opportunities internet radio makes possible. We believe such endeavors can help to facilitate the reconnection of authors' written words and spoken voice—and all the communicative possibilities afforded by the reintroduction of sound in new and exciting ways—and foster the power of "winged words."

PEDAGOGICAL USES OF INTERNET RADIO

This section of our essay outlines ways Dene has harnessed Nouspace Internet Radio in her classes, and it suggests specific applications of internet radio for rhetoric and composition courses.

We should mention that we began the internet radio project during the semester that Dene began hosting a monthly literary event for students interested in reading their own creative works. Once a month on Wednesdays she and local poet Nicki Roseman hosted a program called "Wednesday Literary and Media Arts Café." Using a video camcorder, Dene recorded each performer at the event. She then prepared the audio portion of the recording for radio broadcast and edited the video into clips of individual performers, archiving these resources on our website. The following Sunday evenings leading up to the next Wednesday Café, we broadcasted the audio portion via Nouspace Internet Radio on a program called "Wednesday Café on Sunday Nights," and we provided links to text and video clips of the event from our website (see Figure 14.5). With the recent innovation of the video iPod, these video files are now part of the mobile revolution that began with sound.

At the same time Dene was using Nouspace Internet Radio for literary readings, she was teaching a graduate-level digital media and rhetoric course called "Logomedia, or the Effect of Medium on Literature." It was in that course that she read Aboriginal myths over the radio to her students. The introduction of internet radio into her class pedagogy provided her students with a unique aurality experience—that is, listening via the Web broadcast—and helped them rethink the canon of *delivery* by separating it from its dependence on vision and fixity. Students were relying mostly on hearing and hearing together in a particular space and time. This sense of hearing was emphasized in a particular assignment in which students concentrated on the word as *ephemera* emanating from their computers'

Figure 14.5. Wednesday Café Downloadable Video

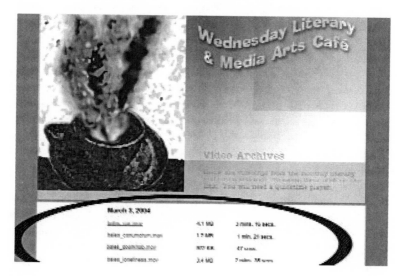

speakers, in what Dene calls the "pneumatic medium," rather than the word inscribed on a "surface medium" like the computer screen or paper (Grigar & Gibson, 2006). Students had to tune in and listen to the spoken word, and they were provided an opportunity to hear the power of words.

Probably the most pedagogically logical use of Nouspace Internet Radio has been its presence in online teaching, specifically in Dene's literary analysis course. This course provides English majors with in-depth training in analyzing fiction, poetry, and drama, and it is required for all students who plan to teach English at the secondary level. Essentially a 14-week boot camp, literary analysis ensures that students are prepared in such areas as genre study, literary history, terminology, prosody, literary theory, and analysis writing. Because of its heavy emphasis on critical thinking, reading, and writing, literary analysis was chosen as an upper-division English course for the new online general studies major. So, in Summer 2004 Dene was asked to create an online iteration of the course for this major. The conversion of the face-to-face class to an online one posed a few challenges. One of the most salient Dene faced when planning the online version was the students' need to *hear* the pronunciation of authors' names, literary terms, and basic vocabulary. Thus, during the orientation class where she discussed Blackboard tools, MOO chats, and blog postings, she also showed students how to download APPLE iTunes, open a stream, and type in the IP address for Nouspace Internet Radio. She also showed them the website for the radio project where audio files would be archived as MP3 files for future listening and download. Then, during the semester, she hosted periodic Sunday night sessions where she offered pronunciation of names and terms. She captured her sessions using ROUGE AMEOBA Audio Hijack, a software program that allows users to record sound while it plays online, and posted them as files for students' future use. Students who could not make a session, who had trouble connecting to Nouspace Internet Radio, or who were using dial-up internet accounts that made streaming audio an arduous task were able to find the files later in the week on the radio project website. Dene also incorporated her enCORE MOO—Nouspace Virtual Environment—in her class. In the MOO, she built a "generic radio object" housing it in her virtual literary analysis classroom and anyone using the room could hear the Nouspace Internet Radio broadcasts (see Figure 14.6). This meant that students using the MOO could chat and click on pertinent links while they were listening to the session. The end result of these broadcasts was that she was able to teach students words like *onomatopoeia*, which many have never seen, much less heard, ever before. Students could learn the correct pronunciation of an author's name, like Milan Kundera or Yann Martel, names they would need to pronounce in their own classrooms one day.

Although these examples have focused on uses of internet radio for literary purposes, there are just as many applications of internet radio for rhetoric and composition classes.

Figure 14.6. Internet Radio Object in Nouspace Virtual Environment

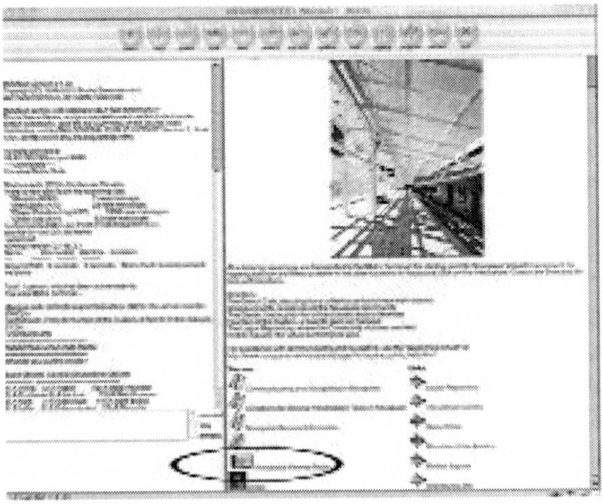

In Dene's graduate rhetoric course, "Studies in Writing for Net Environments"—a project-based course intended to provide Ph.D. students in rhetoric with a theoretical understanding of internet technologies as well as a hands on experience with producing content for the net—she could ask students to create a collaborative radio program. This project would require that students not only deliver their own radio program but also write all the content for it. Through such a project, students would need to become familiar with the radio medium by listening to a range of radio programs already delivered via the Net, such as those produced by National Public Radio. Further, students could learn more about the differences between the formats and delivery of Internet and AM/FM radio. These dimensions of the project would provide students the tools they would need for evaluating their own success and reflecting on their own learning—a process that would certainly entail written documentation. Issues that arise relating to noise, audience, delivery, arrangement, and medium could provide fodder for analysis. In sum, the practical and theoretical possibilities are rich in undertaking a project using internet radio, and a course, such as Dene's, could emerge as a digital media and rhetoric course that explores new media studies, classical rhetoric, and communication arts.

It does not take much more creative energy to envision a similar course for first-year composition students. "Readings" could include much of the literature and media mentioned earlier in this essay: H.G. Wells' *War of the Worlds*; Orson Welles radio program, *War of the Worlds*; the radio broadcast of the Hindenburg accident and its video counterpart; Martin Luther King's "I Have a Dream" speech in print, on video, and the audio version; Aboriginal myths in both print and audio formats; and sound clips of Winston Churchill and Franklin D. Roosevelt; as well as Marshall McLuhan and Quentin Fiore's classic work, *The Medium is the Massage*. These "texts" would raise awareness of differences between genres and media and prepare students to analyze aural contexts and to produce aural content. Additional texts might include Chris Priestman's *Web Radio*, which provides the most detailed information about production, management, and promotion of internet radio; Todd Cochrane's *Podcasting: Do It Yourself Guide*; which lays out the ins and outs of podcasting; and Glenn Fleishman's short essay, "How to Record a Podcast," which offers a quick explanation of recording podcasts. Much of these works could be housed on a class blog. Some blogs, such as Motime.com, even provide space for users to upload files and to provide RSS feeds, making podcasting, and so mobility, an option. In fact, Motime now offers a "mediablog" option which can allow you to publish and organize media files from portable devices like a PDA or cell phone (Motime.com).

Obvious projects include creating radio programs, writing copy for those programs, delivering those programs on the air and as files for podcasting from a class blog, and producing materials for promoting the programs—all or some of which can be accomplished as collaborative assignments. For example, the assignment like the one Dene required with her graduate students involving Aboriginal myths would ask students to research the culture along with oral narratives, identify the kind of material their audience would find engaging, read through the many myths found at the website to identify those that fit that audience, write the content surrounding the program on Aboriginal myths, produce the blog from which the project would be accessible, employ oral presentation skills to broadcast the program effectively, and develop posters, press releases, and other materials for promoting the program. In other words, students would be required to do a lot of thinking, strategizing—and writing.

In addition to these creative, technical, and business communications generated from this experience, there are other contributions that speak to higher-level critical thinking: audience, medium, rhetorical, and marketing analyses and critical and self-reflective evaluation. For example, students in the audience listening to the program could be asked to compare the myths as they appear as written text on the website and as aural text as they *appear* on the broadcast. How different are the two experiences? What drives this difference? What changes need to be made in order to compensate for these differences? "Comparative media," the emerging field interested in this kind of activity, speaks not only to future skills students will need in a world where multimedia information is the norm but also of good, old-fashioned skills in critical thinking. Students involved in the broadcast, on the other hand, could be required to produce a self-reflective evaluation of their work, focusing on issues such as presenting an effective message and meeting expectations of their audience.

Such real-world experiences, as we have outlined, have the potential of enlivening student writing by allowing students to choose their own subjects, providing students with an audience for *hearing* their work (and, perhaps, with blog postings, also *seeing* their writing), and offering a end goal for their work. In essence, a project such as we describe allows for a "student-based approach to writing" (p. 232), suggested by James Middleton and John Reiff (1985). That it constitutes writing created for the Web means that students have the potential of producing more careful work because, as Catherine Smith (2000) points out, students "take real-world writing more seriously when it is done on the Web, where it might actually be seen and used" because their "sense of play and familiarity with online environments . . . stimulate investment in and engagement with their writing" (p. 241). Certainly, work by Margaret Mansfield (1993) has shown, there is a place for real-world writing in a writing classroom (pp. 69-70).

CONCLUSION

In conclusion, with the ascension of the World Wide Web and its emphasis on graphical interfaces, one might argue that the power of the spoken word has been overlooked in the rush to distribute text and images via new electronic contexts where successful communication is often judged as simply the delivery of a message from sender to receiver. Discussions about the higher level of seriousness found in political debates delivered via the radio over debates delivered via television, however, combined with the growing use of wireless technologies for information dissemination, point to the potential of internet radio as a site for public and artistic expression in an increasingly mobile world. That internet radio began as a site of sociopolitical activism is evidence of this potential (Lee, 2005). Yet despite this history, little attention has been paid to sound, much less the rhetorical intent of a message or the way in which an audience would receive and, perhaps, interact with a message. Re-imagining streaming audio content for scholarship and communication in wireless, mobile contexts provides us another opportunity to examine the relationship between orality and aurality, between message and audience, and between visual culture and a growing aural one, a culture that needs tools for understanding the *gravitas* behind the words of any speaker who would attempt to persuade or convince us to believe, accept, or follow a particular concept or action plan.

This reapplication of the layers of rhetorical intent to spoken words adds exciting prospects for multimedia in that it allows additional ways both to examine the intent behind a speaker's words and to encourage a more direct connection between the spoken word and its printed form. In a time increasingly influenced by visual images, being able to hear authors deliver their own words, in the intended rhetorical style, adds greatly to the ability to understand background and intent, both of which can be examined simultaneously in different ways through other forms of media. Finally, in a network-fast media age where increasingly one's image as well as words can be fabricated, the prospect of paying more attention to the pace of the spoken word may offer advantages for contemporary scholarship, especially research seeking to understand the connection between personal *ethos* and cultural change. Projects such as Nouspace Internet Radio can provide the opportunity for the exploration and use of "winged words."

REFERENCES

Affleck, Andy Williams. (2006). *Take control of podcasting on the Mac* [Online]. Tidbits Electronic Publishing.

Asimov, Isaac. (1991). Future fantastic. In Isaac Asimov & Ralph McQuarrie (Eds.), *Robot visions* (pp. 426-431). New York: New American Library.

Beck, Alan. (1999). Is radio blind or invisible?: A call for a wider debate on listening-in. *World Forum for Acoustic Ecology*. Retrieved July 1 2004, from http://interact.uoregon.edu/MediaLit/WFAE/readings/blind.html

Bradbury, Ray. (1992). *The Ray Bradbury chronicles*. New York: Bantam.

Cantril, Hadley. (1940). *The invasion from Mars: A study in the psychology of panic with the complete script of the famous Orson Welles' broadcast.* Princeton, NJ: Princeton University Press.

Fleishman, Glenn. (n.d.). How to record a podcast. *Mac devcenter.com.* Retrieved January 25, 2006, from http://www.macdevcenter.com

Grigar, Dene, & Gibson, Steve. (2005). Ephemeral writing. In *Proceedings of the 2006 Digital Arts and Culture Conference,* Copenhagen, DK, CD-ROM.

Homer. (1996). *The odyssey* (Richard Fagles Trans.). New York: Viking.

Koch, Howard. (1971). *The panic broadcast*. New York: Avon.

Lee, Eric. (2005). *How internet radio can change the world: An activist's handbook.* London: iUniverse.com.

Mansfield, Margaret A. (1993). Real world writing and the English curriculum. *College Composition and Communication, 44,* 69-83.

Middleton, James E., & Reiff, John D. (1985).A 'student-based' approach to writing assignments. *College Composition and Communication, 36,* 232-234.

Motime.com. (n.d.). Motime home page. Retrieved January 25, 2006, from http://www.motime.com

Perkowitz, Sidney. (2004). *Digital people: From bionic humans to androids.* Washington, DC: John Henry.

Radio art: Is there life after death? (2004). *Net Art News.* Retrieved July 2004, from http://rhizome.org/netartnews/index.php

Rogue Ameoba. (2006). You're on the air with...Nicecast. Retrieved April 22, 2006, from http://www.rogueamoeba.com/nicecast/

Shannon-Weaver Model. (2004). *Communication studies, cultural studies.* Retrieved July 1, 2004, from http://www.cultsock.ndirect.co.uk/MUHome/cshtml/introductory/sw.html

Smith, Catherine. (2000). Nobody, which means anybody: Audience on the World Wide Web. In Sibylle Gruber (Ed.), *Weaving a virtual Web: Practical approaches to new information technologies* (pp. 239-249). Urbana, IL: NCTE.

15

DANCING WITH THE IPOD[1]

EXPLORING THE MOBILE LANDSCAPE
OF COMPOSITION STUDIES

Lisa Meloncon Posner
Beth Martin

I could use just a little help . . .
You can't start a fire without a spark . . .
Even if we're dancing in the dark.

Bruce Springsteen

Composition scholars might wonder what dancing with the iPod has to do with writing, but in Bruce Springsteen's words, we compositionists "could all use just a little help." At one time or another, every composition teacher has felt like he or she was "dancing in the dark" in his or her classrooms. Composition scholarship attempts to illuminate new areas of inquiry through spatial metaphors. These metaphors help students and teachers find their own place within the classroom and their writing, making them the starting point for the student's own construction of meaning. Spatial self-awareness in a specific context—even one where difference, alterity, and conflict are encouraged and embraced—allows our students to produce written, discrete texts such as essays, papers, presentations, and websites that are attuned to their own personal and academic spaces. This critical awareness of spatial context provides students the opportunity to demonstrate a self-reflective understanding of place. As technology has continued to impact our students, our classrooms, and ourselves, we must acknowledge changes in our reading, writing, and teaching practices that are spurred by technological shifts. Presently, mobile technologies reframe our geographies of composition theory and pedagogy, taking us into uncharted and little understood territories. Thus, composition scholars and teachers need new ways of envisioning the classroom to better understand the impact of mobile technologies. To that end, we construct a composition paradigm by extending cultural geography's theory of landscape. We use both metaphorical and material constructions of landscape to develop what we call the mobile landscape of the composition classroom. After a brief overview of spatial theory within composition studies, we employ cultural geographer Donald Meinig's notion of successive landscapes to build our own concept of the mobile landscape. We then analyze Apple's iPod as one technological means to enter, explore, and better understand the mobile landscapes in and around the composition classroom. We conclude by pointing to specific strategies to integrate the iPod critically and consciously into the composition classroom, extending the classroom into the mobile landscape.

COMPOSITION STUDIES
AND SPATIAL THEORIES

Geography acted as the support,
the condition of possibility . . .

Michel Foucault

Composition scholars have long used spatial metaphors—outlines, maps, zones, boundaries, cities, sites, locations, margins, architectures—to construct realities of writing. From Mina Shaughnessy's frontiers to Gloria Anzaldua's borderlands, spatial metaphors have provided compositionists with ways to envision, talk about, and create writing spaces. When Shaughnessy (1978) first wrote of frontiers, scholars critiqued her choice because the language reflected a position of colonization (Horner 1992; Lu, 1991). Our scholarship moved away from the frontier metaphor toward communities, cities, and zones, but we continue to engage and critique such spatial metaphors in our discipline. However, relatively little work has been done on the concept of space itself and our persistent use of spatial metaphors. Recently, several composition scholars have begun to address this gap by challenging composition's spatial metaphors and constructing their own spatial contexts for research. As these scholars note, we need to be cautious when using geo-rhetorics, especially in combination with technology, because specific consequences are attached to our metaphorical representations of space.

In her *Geographies of Writing*, Nedra Reynolds (2004) uses postmodern geographic theories to introduce a robust interpretation of spatial metaphors. Reynolds connects cultural geography and composition as she argues for "geographic rhetorics" that study writing and "inquire into the relationships between writers, writing and all places, spaces, sites, and locations" (p. 4). Reynolds contends that the spatial metaphors of frontier, city, and cyberspace have helped to create disciplinary status, but the potential of spatial metaphors, specifically travel and borderlands, have not yet been realized. Reynolds (2004) wants to use the tropes of travel and borderlands as a means "to change our ways of imagining writing through both movement and dwelling—to see writing as a set of spatial practices informed by everyday negotiations of space" (p. 6). She uses geography to connect the material and metaphorical drawing on Edward Soja's trialectics of space, or thirdspace, to "address and resist these containerized conceptions of space" (p. 45). Reynolds advocates for the promise of connecting geography and composition.

Johnathon Mauk (2003) also emphasizes spatial analysis in discussions of composition by drawing on Soja's thirdspace as his "heuristic for orienting the acts of teaching and learning writing in increasingly spaced-out college environments" (p. 370). In a move similar to Reynolds', Mauk uses Soja to connect the spatial and material aspects of students' lives where the "real and imagined" intersect. Employing Soja in this way, Mauk foregrounds the importance of spatial theory to writing and how those spaces impact our teaching practices. Darin Payne (2005) then extends Mauk's investigations into the "*virtual* spatialities that are shaping our student's intellectual lives and reconfiguring the academic landscape" (p. 484) when he addresses how course management systems produce space. Payne provides a useful commentary on spatial theory specifically as it relates to technology, but Payne's examination is limited to the site of content management systems.

Johndan Johnson-Eilola's *Datacloud: Toward a New Theory of Online Work* (2005) provides an even fuller conceptualization of spatial theory and technology. In *Datacloud*, Johnson-Eilola discusses technologies and activities that "at first glance [seem] to be irrelevant," but instead, mark an important cultural moment that moves "toward an understanding of culture and technology as contingent, multi-dimensional, fragmented, and constructed in local uses" (p. 9). He achieves this by showing how we are a "networked culture" where even "apparently trivial activities," such as the use of an iPod, are important to "extend these cultural shifts in productive ways" (p. 9). Using the computer interface as an example, Johnson-Eilola allows for the construction of the interface as a constantly shifting techno-logical space. This conception of space as fluid, contingent, and open to change gives us a way to connect current composition scholarship to cul-tural geography in productive and useful ways.

CULTURAL GEOGRAPHY
AND SUCCESSIVE LANDSCAPES

We are children of our landscape;
it dictates behavior and Even thought
in the measure to which we are responsive to it.

Lawrence Durrell

Geography is important to most disciplines because it is an empirical real-ity. The "where of things" assumes primary importance in humanist dis-cussions, and so it makes sense that compositionists would adopt geogra-phy's vocabulary. Place and space as appropriated by composition scholars demonstrate our ongoing efforts to understand and teach writing process-es. Our descriptions are nothing if not geographical, and they operate in complex interactions among language, society, student, technology, teacher, and classroom. Each of these terms can be paired with another to form a spatial expression of writing, representation, and self. For example, student and society signals how a student's writing process will be affect-ed by her geographic history or where and how she grew up. Alternately, the student's writing process is also impacted by society in terms of where she chooses to write. Composition teachers and scholars continue to inter-pret space and place in composition classrooms, and because of its empha-sis on agency, cultural geography, in particular, provides one means to understand our shifting culture as we migrate to mobile technologies.

Cultural geography is a "sub-field of human geography that focuses on the impact of human culture, both material and non-material, upon the natural environment and the human organization of space" (Cosgrove,

1994, p. 111). Two major branches of cultural geography exist: the Berkeley School, started in the 1920s, and the "new" cultural geography movement, which began in the 1990s. The Berkeley School is founded on the work of Carl Sauer who initiated geography's movement to study human interventions in transforming the surface of the Earth. Sauer (1925) himself provides a succinct description of cultural geography: "The cultural landscape is fashioned from a natural landscape by a culture group. Culture is the agent, the natural area the medium, the cultural landscape the result" (p. 46). During the 1960s and 1970s, cultural geography suffered a decline, but in the early 1990s, a renewed interest in cultural dimensions of geography arose. Driven by different theoretical assumptions on culture difference, cultural geographers started to examine contemporary and urban societies and became interested in nonmaterial culture (e.g., ideology, meanings, attitudes, power, and identities). How these cultural attributes are distributed spatially, and how they relate to the spatial distribution of wealth, power, and justice ground the "new" cultural geography movement. Both branches are still practiced in geography today, and as a subfield, cultural geography continues to grow. This brief discussion, however, points to the difficulty geographers face when trying to define the boundaries of their field; generally cultural geographers admit that the field seems to be "all over the place" (Shurmer-Smith, 1996, p. 2).

Despite geography's own boundary issues, landscape—as the central medium of analysis—continues to be a unifying concept of geography, and it joins different methodologies and theoretical assumptions across the field. In particular, cultural geographers center much of their scholarship on landscape because it "represent[s] an attempt to define the elusive object of geographical study in an integrated yet more focused way" (Matthews & Herbert, 2004, p. 217). This promise of integration is primarily why we borrow the term "landscape" for our analysis of mobile technologies in composition studies. Let us make clear what we mean by landscape. "Landscape denotes the interaction of people and place: a social group and its spaces," declares Paul Groth (1997), "particularly the spaces to which the group belongs and from which its members derive some part of their shared identity and meaning" (p. 1). Adopting Groth's definition, we emphasize landscape as material space mediated through culture representations. This means that landscape is socially constructed, uniquely interpreted by individuals and groups. Meaning in a landscape is negotiated through an individual's or group's attempts at spatial exploration, which merges the physical materiality of that space with our social and cultural constructions of it. Landscapes are often defined in terms of location and boundaries, but they also can be defined in terms of context. The contextual nature of landscapes is affirmed in the work of cultural geographer Donald W. Meinig.

Meinig (1979) believes a landscape is "composed of not only what lies before our eyes but what lies within our heads" (p. 34), thus accounting for various perceptions of the "same" landscape. When looking at a particular scene, viewers incorporate their own beliefs, values, hopes, and fears into their assessments of the landscape. Each viewer "sees" the scene through a different lens: nature, artifact, system, problem, wealth, ideology, history, place, and aesthetic (pp. 35-46). To be able to "write" the descriptions, the viewer has to be able to "read" the landscape. Meinig speculates that the viewer socially constructs his or her description of a landscape, and he asserts that the description "tell[s] us much about the values we hold and at the same time affect the quality of the lives we lead" (Meinig, 1979, p. 47). The different lenses put forth by Meinig are only a fraction of the lenses that can be employed when viewing landscape. What makes his theory of "versions of the same scene," or successive landscapes, important for geography and for compositionists is that it brings together the two schools of cultural geography by making room for unity and cohesion (the Berkeley School) and postmodern difference ("new" geography). In this sense Meinig provides two important bridges: first, his theory allows for multiple interpretations of landscapes, and second, it directly connects geography to composition by encouraging a complex "reading" of landscapes.

Part of cultural geography's historical evolution meant it "turn[ed] away from *explanation* and towards an exploration of *meaning*" (Mitchell, 2000, p. 64, emphasis in original). Because compositionists too are invested in understanding meaning, we share a common investment with geographers. A large part of the move from explanation to meaning in geography relies on exploration and "reading" the landscape. In many ways this technique mirrors the long-standing tradition in composition of critical analysis through close examination of texts. Meinig's successive landscape theory calls for the application of different lenses to develop a complex meaning of the space we are examining. For example, if you were to "read" this essay as a successive landscape, you apply a lens, perhaps the aesthetic, and then another, and another. When taken all together, each of these "versions of the same scene" becomes a successive landscape with multiple ways of "seeing" the same text. By combining the lenses to create a successive landscape, a reader—or writer—creates a more complex, thicker description/production of the space of a text. Successive landscapes also allow us to understand the relationship between individual and cultural interpretation because they are extensions of ourselves and of the spaces we inhabit. Further, Meinig's theory can help us understand the imbricated, complex, and culturally constructed mobile landscape.

As we move through our lives in a complex series of spatial relationships, mobile landscapes are those landscapes that we move with us and that are also affected by our movement. In this sense, mobile landscapes encapsulate the mobility that is an "unending patient readjustment to circumstance" (Jackson, 1984, p. 151). The mobile landscape is differentiated from other landscapes by its multiple lens and attention to individual, flexible constructions. The mobile landscape affords the teacher and student the opportunity to create, view, compose, construct, and participate in learning environments that are portable, interactive, and individually created to address the diversity of the students and to suit the pedagogical aims of the classroom. Mobile landscapes provide new opportunities for writing instruction because they incorporate different "versions of the same scene." Incorporating a mobile successive landscape into our interpretations and "reading" of texts and spaces, teachers, scholars, and students have the opportunity to develop expanded practices for how, where, and what we write.

If the landscape is mobile—always shifting and changing—how do students and teachers define the place of the mobile landscape? Mobile landscape creates a unique problem, one that can be solved by combining Meinig's theory of successive landscapes with the technological tools of mobility. To be able to "read" and "write" the mobile landscape, then, students must negotiate seemingly endless choices as to their entry point—the technology—and their movement—use of technologies. In our case, we have chosen the Apple iPod as one technology in the mobile landscape. The iPod allows students to engage the mobile landscape, moving through material and metaphorical spaces. In the upcoming sections, we will investigate the iPod as part of a mobile landscape and through the lenses of technology device, social phenomenon, and educational tool.

IPOD AS PART OF A MOBILE LANDSCAPE

She blinded me with science
And hit me with technology

Thomas Dolby

Kathleen Blake Yancey (2004) calls for developing a new curriculum of rhetoric and composition that specifically addresses circulation and delivery (p. 311). The mobile landscape can play a key role in Yancey's framework because it offers the composition teacher and student the means to consider both the medium of circulation and the technology of delivery. The mobile landscape is important to our students both symbolically and materially, and this landscape is embedded in the social spaces of our classrooms. The binary relationship between the "in here" of the classroom and "out there" of the "real world" needs to be deconstructed. As writing teachers, we are positioned to disrupt this binary by connecting and complicating the inside-outside aspects of our students' literate practices. The iPod is one technology that may help us in our critique of this binary. Students bring certain technologies into our writing classrooms simply because they are part of their lives. By harnessing "their" technologies, such as the iPod, for use in writing classrooms, we can begin to merge technologies, practices, and literacies of the mobile landscape. This act of merging helps us create a richly layered, interactive pedagogical strategy that engages students with varied learning styles and enhances a compositionist's effectiveness. Toward this more complicated view of the mobile landscape, we turn our attention to the technological, social, and educational scenes of the iPod. We acknowledge that these scenes are overlapping, but we attend to them individually to emphasize the unique way each contributes to the mobile landscape. Through these three scenes, we are creating our own version of a successive mobile landscape.

Technological Practices of the Ipod

Here we are now. Entertain us.

Nirvana

According to Apple's description (2006a), the iPod has a 2-inch (diagonal) grayscale LCD with LED backlight, 12 hours of battery life, and USB 2.0 and FireWire 400 connections. It weighs 5.6 ounces and boasts a 20, 40, or 60 gigabyte hard drive. The iPod comes with earbud headphones, AC adapter, and a FireWire cable, all of which are iPod white. The iPod and its iTunes software work with Apple and IBM compatible music players. The iPod's wireless features include home and car stereo connections as well as remote navigation capabilities (¶ 1-2, 4). With the ability to store up to 5,000 songs the iPod is primarily known as a well-engineered music delivery device, but Apple clearly states that this device is more than a music player. Apple tells its college student buyers that the iPod's benefits include:

integrates with other technologies to create slideshows, soundtracks, and movies; makes projects "come to life" by including digital photos; transfers text–based information so notes, articles, and "flash cards" can be "right at hand"; keeps track of your schedule and addresses; and loads and transfers files with ease (¶ 3-7). The iPod's many versions combine to make it the most popular music player on the market today (*Business Week Online*, 2005, ¶ 1).

The iPod is also an aesthetically pleasing, beautifully engineered piece of technology. Recent introductions of new iPod products such as the "Nano" include television and video viewing capabilities, podcasting, photo storage, and even cell phone integration, and Apple continues to hype the iPod's potential to offer new, extended technological capabilities. The experience of using the iPod is different from other mobile technologies such as laptops, PDAs, and cell phones. The iPod's design and portability lend to its technological efficiency and its "cool" factor. Its design and size makes it faster and easier to use for video viewing and audio listening than a laptop. And although PDAs are primarily designed to be professional organizers, their storage capacity can be significantly smaller than an iPod's. Some PDAs are not specifically intended to play audio files, although some now try with limited success. Mobile phones require monthly payments for video and audio services, and these payments may be prohibitive for students. iPod users also need not worry about being in the "right area" to receive a signal. When the iPod battery is charged, it is ready to go anywhere. The iPod is highly portable and can be used in automobiles through wired and wireless technologies. In addition, a new patent has been filed to "un-wire" the iPod even more by creating a WiPod, which is wireless iPod (Fadell, Zadesky, & Filson, 2004). All of these iPod features open up additional dimensions of the mobile landscape as students can listen to— or perhaps even view—lectures while driving, riding, or walking across campus. The iPod frees the composition teacher from a single location for learning to an on-demand learning environment, which then may help fully realize learning as part of the mobile landscape. The iPod, or other technological devices, are paradoxical in that they allow for mobility within spaces and temporary stability through the creation of spaces. For example, the iPod is a transportable device for music listening, and in some cases, music and voice recording.

The iPod was designed primarily as a source of entertainment, which seemingly renders all other uses secondary. These secondary uses, however, should be acknowledged in scholarly discussions because the device can be employed as a substantive tool for learning and communication. The iPod is a "visible" tool. Unlike word processing programs that have become "invisible" because of their persistent use, the iPod currently relies on visibility as part of its marketing strategy. This visibility creates the potential for anytime, anywhere access to learning: The iPod is a technology students will *want* to use. We see the iPod's visibility as one means to encourage students to use the technology as an extension of classroom practices. With the iPod our students would be able to re-invent the process of peer critique through audio feedback or listen to instructor revision comments as they are revising. The technological capabilities of the iPod afford teachers and students the means to transgress the boundaries of both the classroom space and writing practices. Writing then becomes part of the mobile landscape that our students occupy in their everyday lives.

Cynthia L. Selfe (1999) warns of the danger posed when technology becomes invisible because it allows us to ignore the consequences of using classroom technologies (p. 411). If the iPod becomes yet another invisible technology—so ubiquitous that it is normalized, we as scholars may deny the critical dimensions of this technology in the lives of our students. The iPod is part of most students' technological literacy, and as Selfe (1999) explains, technological literacy is necessary for our students to "communicate as informed thinkers and citizens in an increasingly technological world" (p. 414). Because of its visibility, the iPod can be a desired literacy for students, thus making it easier to integrate it into other avenues of teaching, but it also becomes paramount to prompt students to think critically about technology by developing a "technological literacy" beyond just knowing what buttons to push. This literacy is more akin to Stuart Selber's (2004) "multiliteracies for a digital age" in which technological applications involve "values, interpretations, contingency, persuasion, communication, deliberation, and more" (p. 235). We must extend the use of the iPod as a mode of critical literacy constructing the mobile landscape.

Social Dimensions of the Ipod

Gimme fuel, Gimme fire, Gimme that which I desire.

Metallica

Technological phenomena do not act in a vacuum, and the social implications of iPod are many. The iPod, as a tool, provides entrance into a specific discursive community. Its white "earbuds" signal membership in the iPod community of cool. Students can download celebrity "playlists" and inculcate themselves, even more, into the culture of celebrity. This "cool" factor comes in two forms: the productive and positive aspects of this social phenomenon and the limiting aspects categorized by socioeconomic class.

As social phenomenon, iPod has not only secured its place as market leader in technology but also spawned an entire secondary market of products, which extends the reach of the iPod's "coolness" factor. Secondary markets of accessories, add-ons, and iPod specific products extend the product's mobility in and among a variety of different mobile landscapes. For example, one popular accessory is Hasbro's i-Dog. The i-Dog works like a speaker, grounding the mobility of the iPod to a specific, stationary location. However, the i-Dog itself moves and blinks when music is played, possibly reminding listeners of the iPod's mobile possibilities. Apple iTunes and podcasting are creating a social network of information exchange. iTunes is a media player for playing and organizing digital music and an interface for buying music. iTunes commercials feature silhouettes dancing alone—making space for themselves through individual expression and as part of the larger iPod community. It is this type of mobile individuality rooted in a larger community that offers composition teachers the impetus for iPod use in the classroom. iTunes also allows for the purchase of podcasts. Podcasting combines the "Pod" from iPod with the "casting" from broadcasting. Quite literally, a podcast is either audio or video files in the form of RSS or Atom feeds delivered through the Internet. These files can be downloaded to play on a computer or through a mobile device such as an iPod, and some podcasts can be automatically updated through a subscription. Podcasting highlights iPod's potential as a learning device through its productive possibilities. For example, students can create podcasts throughout the course of the semester, and these podcasts can be stored for use in future classes. Interactive lectures and classroom discussions can be uploaded for reference. Podcasting becomes a vital aspect of the mobile landscape because it renegotiates the boundaries of inside-outside the classroom. The classroom becomes portable, mobile, and on-demand. The iTunes store even contains a category for free downloads of "courses and lectures."

The "cool factor," entertainment value, and knowledge creation and dissemination of the iPod are connected to socioeconomic issues. The high-end iPod retails for $299.00, and although the iPod shuffle costs less at $99.00, it does not offer the same features and storage capacity of its more expensive counterpart. These costs, however, are but one dimension of the socioeconomic factors of the iPod. We need to be fully aware of the class consciousness of our educational devices, for as Charles Moran (1999) cautions, " the study of technology needs to grounded in the material as

well as the pedagogical and cultural and the cognitive if it is to be intellectually and ethically respectable" (p. 206). Right now, the iPod excludes
large numbers of users from its community, and that community is still
largely made up of consumers rather than producers of the iPod technology. Further, the iTunes website, which may be a model for other interfaces,
is covered with famous and wealthy iPod users who can afford the technology—furthering exaggerating the sense of the haves and have nots. These
socioeconomic concerns reveal that the technology is not just about the
device but also user practices. All students must be given the opportunity
to navigate through various information conduits, such as the iPod, not just
financially secure students.

As we teach in the mobile landscape, we must contend with issues of
technology access. This requires the commitment by educators and their
institutions to provide the necessary technology—such as an iPod— to all
students. This access helps to move students into positions of power, to
provide a starting point for their understanding of information, and to allow
them to construct and distribute knowledge. By using the iPod as a field-
recording tool in and around campus, students can become aware of social
status both metaphorically and materially. The inside-outside binary continues to shift and blur as students move in between the mobile landscapes
of the classroom and the world that surrounds them. Mobile landscapes
offer students and teachers social interactivity that can be realized through
creative, pedagogically driven assignments.

EDUCATIONAL PRACTICES OF THE IPOD

Too much information running through my brain
Too much information driving me insane

The Police

The iPod as part of an educational practice exploits the benefits and applications of the mobility. Universities, both public and private, are already
integrating the iPod in the classroom. According to Apple's web site
(2006c), Georgia College and State University (GC&SU) is one university
working to employ the iPod in the classroom: "Professors at Georgia
College & State University have discovered that the iPod isn't just the
world's coolest device for storing and playing music tracks. Now, students

in several classes tote an iPod around campus to listen to digital audio content that ranges from Shakespeare to Spanish history" (¶ 1). The idea that a "cool" entertainment device is an important pedagogical tool is evidenced by GC&SU's adoption of an interdisciplinary approach in some of its "iPod" classes. In one such class, "War, Politics, and Shakespeare," the iPod was used to combine literature, music, architecture, and art. In this case, faculty members worked together to create a curriculum for the iPod, helping to foster a sense of community among different disciplines and making for a unique classroom experience. Teachers can use the iPod to keep our students involved and encourage them to explore the multiple mobile landscapes available to them. Accessing a curricular mobile landscape helps to not only challenge the inside-outside binary but also erase disciplinary boundaries as well.

In the fall of 2004, Duke University started its now-famous pilot program to incorporate the iPod into the college classroom. All incoming freshman were given an iPod equipped with a voice recorder. Because it was a university-wide initiative, students and faculty engaged in the pilot were from a wide range of disciplines. Courses such as "Body Works: Medicine, Technology, and the Body in Early 21st Century America," "Fullness of Being," and "Economic Principles," all were courses taught offered in the spring of 2005 (Center for Instructional Technology, 2005, ¶ 2). The diverse courses and various ways the iPod was integrated give insight into its potential as an education tool. The iPod has been and can be used to create "different versions" of the mobile landscape. By using traditional methods such as peer critiques and then relating them to the recording capabilities of the iPod, a new version of the "same scene" is realized as the peer critique can occur anytime and anywhere, which allows students to participate in the classroom while physically inhabiting another landscape.

In June 2005, Duke University's Center for Instructional Technology (CIT) released a report evaluating academic iPod integration. The CIT determined the following benefits: convenience for faculty and students by reducing dependence on physical materials; flexible and location independent access to course materials; effective and easy-to-use recording capabilities; greater student engagement in class discussions; and enhanced support for individual learning preferences (p. 2). The report also highlighted institutional impacts of the iPod project, and one of the significant findings was "catalyzing conversations about the best role for technology in teaching" (p. 11). This particular finding intersects with the role of mobile technologies in composition classrooms. Whether the iPod or some other technological device is used, conversations are started where teachers and students can better articulate their needs and concerns about technologies, including a range of technologies as part of the mobile landscape.

Critical analysis of "entertainment" technologies is necessary in our culture. We would be foolish as composition teachers and scholars to believe that the iPod and other devices are simply an "outside" force that we need not concern ourselves with. The forces of these entertainment technologies make lasting impacts on our students and their interactions in our classrooms. How can the iPod be used as a pedagogical device by teachers who already feel overwhelmed by technologies? No single nor simple answer exists to this question, but as Andrew Feenberg (1999) tells us, "technology is the medium of daily life" (p. i), and so we need to be aware of the medium and all that it entails. Composition teachers need to consider the ways in which the iPod and other devices of the mobile landscape refigure the literate practices of our culture. Technologies, specifically mobile technologies, offer much hope and possibility, but with these potentials come cautions. We must share and interrogate our iPod pedagogies. Consider, for example, the use of the iPod as another way to deliver information, which is one aspect of composition pedagogy. For a number of years teachers have been able to deliver information via websites. The iPod can take this one step further by creating an information space that is aural and portable. The iPod also affords teachers the opportunity to create a mobile pedagogy, delivering information in a variety of forms and at a range of sites to allow students opportunities to investigate scenes in that moment. The iPod's memory and portability make it a useful storage device for student research.

Many first-year composition classes are based on argument, and the mobile landscape can allow for a multiple entry points into a single issue. By providing additional access points, mobile technologies can offer students and teachers an expanded approach to composing multifaceted arguments. However, as Meinig cautions, we need to be aware of differences, differences based on our own individual points of access. Thus, difference becomes a focus of our exploration of the mobile landscape. For example, students can use the iPod to interview people who have different views on a particular subject, or students can gather a collection of images and create a corresponding soundtrack that explains multiple sides of an issue. Through techniques such as these, students can present their work to the class. These types of multigenred "compositions" foster advanced critical literacies within the classroom and reflect the technological practices of the iPod within the mobile landscape.

The iPod can serve as an information literacy device in that it provides another avenue for finding, retrieving, storing, and disseminating information. As we move through the mobile landscape, we must understand how the information is changed as the scenes of the landscape change. The iPod draws attention to our "teacher talk"—the way we speak with our students and the ways they listen and engage us. As teachers of language, we already understand how information may differ from text to speech and our understanding will help craft pedagogies that use the iPod beyond its entertainment scene. Students can now pause, fast-forward, and rewind various classroom dialogues thus enhancing the classroom experience. In short, we can add another dimension to information retrieval and information transmittal, and we can allow our students to create their own classroom experience and their own textuality. We can, in the constructed voice of Myka Vielstemmig (1999), "ask students to navigate among all these textualities—not just in print and online, but in talk as well" (p. 111) to engage the mobile landscape.

Podcasts are not only ways for the social aspects of the device to be realized, but educational ones as well. Podcasts create a grounded space for the students' production of audio "texts." The productive possibilities will allow students to explore mobile landscapes and relate their findings back to the class. Also, consider having students use the iPod for scene analysis and field research for their composition classes. The iPod has already transformed the model of music delivery and is now beginning to enter the television and movie industry as well, which means there is little reason to believe that it will not help transform the classroom and the delivery of our content. Benefits of learning enhanced through mobile technologies, such as the iPod, include portability, interactivity, and individuality within specific contexts. In this sense, mobile learning becomes one of the successive landscapes of an educational experience and provides students with the cognitive skills necessary to function in a wireless world.

If we look at our work, as a microcosm of what the students will encounter as they leave the academy, the "necessity" for this type of network is obvious. Technology allows us to navigate our extended networks, and when used in a learning environment, technology serves as a path to knowledge construction. Paul Groth (1997) tells us good landscape analysis is "both seeing and thinking" (p. 16), and we feel that mobile landscape analysis is "seeing and thinking" and hearing and composing. By incorporating varying techniques into the classroom, all types of learners have a better chance to access the mobile landscape.

LET THE MUSIC PLAY

I say, we can go where we want to
A place where they will never find
And we can act like we come from out of this world
Leave the real one far behind
And we can dance

Men without Hats

Our exploration of the mobile landscape and the iPod begins a new chapter of the technology challenges of the 21st century. Geoffrey Sirc (2002) works at the forefront of challenges—his gangsta-centric composition class uses rap music. Sirc's melding of writing and sound clearly works at the intersection of old and new technologies. Imagine sitting in his class listening to rap songs or rap excerpts on an iPod and then discussing and writing about them in class. Sirc's composition as a happening advocates ". . . a basic awareness of how to use language and information, a cool project, and a sense of poetry" (p. 277) and the opportunity for composition teachers to create "dwellable, exciting places to inhabit in our curriculum" (p. 216). The iPod can help enable this interactive, student-centered curriculum. The iPod may even help realize Sirc's idea that "hip hop is a rubric for some of the most exciting cultural media available to your people today, transcending perceived distinctions of age, gender, race, and ethnicity" (p. 271). Composition as a happening also argues for movement and mobility. As part of this mobility, composition teachers can see the mobile landscape, dwell in it, and realize the benefits—the opportunities for our students to engage with and live through their writing. The iPod as a tool for analysis can also be a tool for invention and production via podcasting and integrating recordings in new media compositions.

The iPod is just one device in the mobile landscape, and as Meinig argues, landscapes are synonymous with our lives. The making of the mobile landscape can help us imagine new possibilities, examine who we are in the culture, and try on various roles as we negotiate in/through the mobile landscape. Pierce Lewis (1979) posits ". . . landscape is our unwitting autobiography, reflecting our tastes, our values, our aspirations, and even our fears in tangible, visible form" (p. 47). We are presently building a mobile landscape for composition, and it is promising and exciting, as well as frightening and risky. With the advent of mobile devices, we cannot ignore that our field will transgress boundaries of inside-outside our class-rooms. There are too many options to consider and too many forces at work for compositionists to teach writing in a single way. The inclusion of the iPod and other mobile devices into the classroom curriculum goes toward Sirc's emphasis on composition as a "happening." The iPod, as a conduit for literacies, shapes our students thinking one byte, one sound, and one playlist at a time, and as Yi-Fu Tuan (1979) explains "[l]andscape allows us and even encourages us to dream" (p. 101). To compositionists and students, we say, "Let the music play!"

ENDNOTE

1. iPod is a registered trademark of Apple Corporation.

REFERENCES

Apple Corporation. (2006a). Apple-iPod family. Retrieved January 14, 2006, from http://www.apple.com/ipod.ipod.html

Apple Corporation. (2006b). iPod in the classroom. Retrieved January 14, 2006, from http://www.apple.com/education/ipod/

Apple Corporation. (2006c). A pocketful of learning. Retrieved January 14, 2006, from http://www.apple.com/education/profiles/georgia/

Business Week Online. (2005). MP3 market share. Retrieved January 14, 2006, from http://www.businessweek.com/technology/tech_stats/mp3s051122.htm

Center for Instructional Technology Duke University. (2005). Duke University ipod first-year experience final evaluation report. Durham, NC: Duke University.

Cobain, Kurt, Novoselic, Krist, & Grohl, Dave. (1991). Smells like teen spirit. Lyrics. Perf. Nirvana. *Nevermind*. Geffen Records.

Cosgrove, Denis. (1994). Cultural geography. In R.J. Johnston, Derek Gregory, & David M. Smith (Eds.), *The dictionary of human geography* (3rd ed., p. 111). Oxford, England: Basil Blackwell.

Dolby, Thomas, (1982). She blinded me with science. Lyrics. *The Golden Age of Wireless*. Venice in Peril, Capitol-EMI.

Doroschuk, Ivan. (1982). The safety dance. Lyrics. Perf. Men Without Hats. *Safety Dance*. MCA Records.

Durrell, Lawrence. (1961). *Justine*. New York: Dutton.

Fadell, Anthony M., Zadesky, Stephen Paul, & Filson, John Benjamin. (2004). U.S. Patent Pending Application No. 20040224638. Washington, DC: U.S. Patent and Trademark Office.

Feenberg, Andrew. (1999). *Questioning technology*. London: Routledge.

Foucault, Michel. (1977). *Power/knowledge: Selected interviews and other writings 1972-1977* (Colin Gordon, Ed.). New York: Pantheon Books.

Groth, Paul. (1997). Frameworks for cultural landscape study. In Paul Groth & Todd W. Bressi (Eds.), *Understanding ordinary landscapes* (pp. 1-21). New Haven, CT: Yale University Press.

Hammet, Kirk, Hetfield, James, & Ulrich, Lars. (1997). Fuel. Lyrics. Perf. Metallica. *Reload*. Elektra/Wea.

Horner, Bruce. (1992). Rethinking the "sociality" of error: Teaching editing as negotiation. *Rhetoric Review, 11*, 172-199.

Jackson, J. B. (1984). *Discovering the vernacular landscape*. New Haven, CT: Yale University Press.

Johnson-Eilola, Johndan. (2005). *Datacloud: Toward a new theory of online work*. Cresskill, NJ: Hampton Press.

Lewis, Pierce. (1979). Axioms for reading the landscape: Some guides to the American scene. In Donald W. Meinig (Ed.), *The interpretations of ordinary landscapes: Geographical essays* (pp. 11-32). Oxford, England: Oxford University Press.

Lu, Min-Zhan. (1991). Redefining the legacy of Mina Shaughnessy: A critique of the politics of linguistic innocence. *Journal of Basic Writing, 10*(1), 26-39.

Matthews, James A., & Herbert, David T. (2004). Introduction: Landscape the face of geography. In J.A. Matthews & D.T. Herbert (Eds.), *Unifying geography: Common heritage, shared future* (pp. 217-223). New York: Routledge.

Mauk, Johnathon. (2003). Location, location, location: The "real" (e)states of being, writing, and thinking in composition. *College English, 65*, 368-388.

Meinig, Donald. (Ed.). (1979). The beholding eye ten versions of the same scene. In *The interpretations of ordinary landscapes: Geographical essays* (pp. 33-48). Oxford, England: Oxford University Press.

Mitchell, Don. (2000). *Cultural geography: A critical introduction*. Oxford, England: Blackwell Publishers.

Moran, Charles. (1999). Access: The a-word in technology studies. In Gail E. Hawisher & Cynthia L. Selfe (Eds.), *Passions, pedagogies, and 21st century technologies* (pp. 205-220). Logan: Utah State University Press.

Payne, Darin. (2005). English studies in Levittown: Rhetorics of space and technology in course-management software. *College English, 67*, 483-507.

Reynolds, Nedra. (2004). *Geographies of writing: Inhabiting places and encountering difference*. Carbondale: Southern Illinois University Press.

Sauer, Carl O. (1925). The morphology of landscape. *University of California Publications in Geography, 2*(2), 19-54.

Selber, Stuart. (2004). *Multiliteracies for a digital age*. Carbondale: Southern Illinois University Press.

Selfe, Cynthia L. (1999). Technology and literacy: A story about the perils of not paying attention. *College Composition and Communication, 50,* 411-436.

Shaughnessy, Mina. (1977). *Errors and expectations.* New York: Oxford University Press.

Shurmer-Smith, Pamela. (Ed.). (1996). *All over the place: Postgraduate work in social and cultural geography.* Portsmouth, UK: University of Portsmouth.

Sirc, Geoffrey. (2002). *English composition as a happening.* Logan, Utah State University Press.

Springsteen, Bruce. (1984). Dancin' in the dark. Lyrics. *Born in the USA.* Columbia.

Sting. (1981). Too much information. Lyrics. Perf. The Police. *Ghost in the Machine.* A&M Records.

Tuan, Yi-Fu. (1979). Thought and landscape: The eye and the mind's eye. In Donald W. Meinig (Ed.), *The interpretations of ordinary landscapes: Geographical essays* (pp. 89-101). Oxford, England: Oxford University Press.

Vielstemmig, Myka. (1999). Petals on a wet black bough: Textuality, collaboration, and the new essay. In Gail E. Hawisher & Cynthia L. Selfe (Eds.), *Passions, pedagogies, and 21st century technologies* (pp. 89-114). Logan: Utah State University Press.

Yancey, Kathleen Blake. (2004). Made not only in words: Composition in a new key. *College Composition and Communication, 56,* 297-328.

Appendix

16

TERMS FOR GOING WIRELESS

AN ACCOUNT OF WIRELESS AND MOBILE TECHNOLOGY FOR
COMPOSITION TEACHERS AND SCHOLARS

David Menchaca

The terms *wireless* and *mobile*, as used in the phrase *wireless and mobile technology*, refer to a user's ability to roam while using her technologies. Indeed, wireless and mobile technologies are now so thoroughly intertwined that users rarely think of their laptops and cell phones as being either one *or* the other: wireless or mobile. In the realm of personal technology, these two terms have become one. Of course, a mobile phone is also a wireless phone, and most wireless laptops are mobile as well. But this conjoining of terms is not always the case with mobile and wireless technologies. Early Personal Digital Assistants (PDAs) were similar in size to a small paperback book, easily carried and easily operated on-the-go but synchronized with a desktop computer via a cradle connected to the desktop by wires. Many desktops also have wireless capabilities built in to connect with peripherals such as a print server or even a surround sound receiver. However, carrying your desktop's Central Processing Unit (CPU), monitor, keyboard, and mouse would be very difficult, and these systems still need to be plugged into an outlet for power. When the two terms become seemingly interchangeable in common parlance, what exactly do we mean by wireless and mobile technologies? And what have we lost or gained in the coupling and near conflation of *wireless* and *mobile* into *wireless and mobile*?

Identifying how the terms wireless and mobile have changed over time, resulting in their current conflation, is important for teachers and researchers of composition. As Kenneth Burke (1973) notes, "[t]he mere act of naming an object or situation decrees that it is to be singled out as such-and-such rather than as something-other" (p. 4). Naming a technological artifact "wireless" and "mobile" is much different from naming the same artifact "wireless and mobile." "Wireless" and "mobile" maintains a distinction between the artifact's component technologies (i.e., wireless technology and mobile technology). "Wireless and mobile" makes the artifact into a single technology. Thus, naming an artifact "wireless and mobile" necessitates that both technologies are present in the artifact, but further it implies that the removal of either the wireless technology or the mobile technology would render the artifact inadequate for its intended purposes. Changing the terms of a field changes the field itself. "Even if any given terminology is a *reflection* of reality," posits Burke (1969), "by its very nature as a terminology it must be a *selection* of reality; and to this extent it must also function as a *deflection* of reality" (p. 45). The current conflation of the terms wireless and mobile elides opportunities for composition teaching and research: This conflation conceals the histories of the technologies and closes any openings that composition teachers and researchers could use to promote alternate and potentially more productive definitions of wireless and mobile technologies. In other words, our composition pedagogies and research methods are delimited to being "wireless and mobile" if we uncritically accept the conflation of these terms. The implications of this acquiescence are that our pedagogies and research methods are rendered inadequate if either the wireless or mobile component is absent or removed.

By uncritically embracing the conjoined terms *wireless and mobile* rather than maintaining the distinct terms *wireless* and *mobile,* we inadvertently accept a version of reality that may exclude us, or our students. In the wireless and mobile technology industry, the terms are defined by organizations such as the Institute of Electrical and Electronics Engineers, Inc. (IEEE), the American National Standards Institute (ANSI), and the European Telecommunications Standards Institute (ETSI); by commercial enterprises such as Netgear, Inc. and Cisco Systems, Inc.; and by scientists and engineers from a host of disciplines. Conspicuously absent from this list are educators, social scientists, and humanities faculty. The use of wireless and mobile technologies in our composition classrooms and the research that informs that use should begin with an interrogation of the ideologies that help to define wireless and mobile technologies. As teachers and scholars of rhetoric and composition, we must critically investigate these terms as we appropriate them in our pedagogies and methodologies. I do not intend, however, to present this critique, once and for all, here. Instead, I offer a brief history of *wireless* and *mobile* to demonstrate how these two terms came together as part of the ubiquitous coupling of *wireless and mobile technology*. Then, as a teacher and scholar of rhetoric and composition, I offer strategies to recontextualize the terms wireless and mobile to challenge their conflation. Finally, recognizing that wireless and mobile technologies are a complex of written standards, technological hardware, and even user practices, I provide definitions for a handful of terms related to both wireless and mobile technologies. My list is by no means exhaustive, but it helps to open up a conversation on the role of defining technologies and the ways rhetoric and composition teachers and scholars can enter a dialogue on such technologies.

WIRELESS: A BRIEF HISTORY

Translated from its Greek roots, telegraphy means (tele/τελοζ) at a distance and (graphy/γραφη) writing. In general, a telegraph is a technology for sending messages across great distances. The first telegraphic technology to actually be named as such was Chappé's giant semaphore towers.[1] However, long-distance communications technologies have existed since long before the ancient Greeks. Beacon fires communicated a rallying point to soldiers or the contour of the coastline to sailors. Smoke signals have been carrying messages since humans learned to harness fire. Horse and rider and courier pigeon have moved written messages by hoof and wing. We anachronistically know these systems of communication as telegraphic, yet none of them relied on wires to transmit messages. Why then are they not referred to as wireless technologies? The simple answer is that, although they are considered telegraphic systems of communication, they came before Morse's Telegraph and its single, copper wire. The telegraph revolutionized long-distance communication by sending messages faster than a horse and rider, a courier pigeon, or a locomotive, for that matter. In addition, long-distance communication no longer required line-of-sight as with the beacon fire, smoke signal, or Chappé's semaphore. The telegraph rendered these communication technologies obsolete in the industrialized West and added the term *wire* permanently to the common parlance. The addition of the term *wire* changes our perception of long-distance communication technologies as objects of discourse within the technological, social, political, economic, and cultural domains that encounter them. Michel Foucault (1972) explains that the object of discourse is replete with a set of relations that are "established between institutions, economic and social processes, behavioral patterns, systems of norms, techniques, types of classification, modes of characterization" (p. 45). These relations constitute a new discipline—both a field of study and a form of control. That is, by using terms to define an object and bring it into discourse, scholars and practitioners establish a discipline, engage in research, and make knowledge. In other words, wireless technologies, such as PDAs, laptop computers, and cell phones, are wireless because they are technologies capable of long-distance communication that originates in a discipline whose relations are formed in conjunction with and in opposition to the term *wire*. Of course, they are also wireless technologies because they transmit messages via a signal that travels through the air. However, I contend that they would not be known as wireless technologies had the telegraph wires never been strung across the land.

Wired Telegraphy

Samuel Morse is credited with founding the telegraph industry by winning over the U.S. Congress through his 1844 demonstration of the Morse Telegraph, a long-distance communication system using electromagnetism to send a series of dots and dashes from one end of a copper wire to another. Morse's Telegraph represents "the wire" from which modern wireless communication devices are unplugged. The telegraph was a success as a long-distance communication technology, but it was limited to a single transmission at a time. Alexander Graham Bell's multiplexer—a device that varied frequencies to send separate messages at the same time across a single wire—attempted to resolve the problem of the single message transmission of the telegraph. Although the multiplexer provided the technological means to send simultaneous messages, it was not able to satiate the public's burgeoning desire to send and receive messages. Wired communication was bound by material, labor, geographic, time, spatial, and human constraints.

Wireless Telegraphy

Guglielmo Marconi and his radio transmitter and receiver are the technologies that removed the wires from long-distance communication. Wireless technologies still use electromagnetism to send their signals, but instead of passing those signals through a copper wire, wireless technologies use radio waves to transmit their signals through the air, across space and time. An important aspect of radio waves relevant to the field of wireless and mobile technologies is frequency, or the number of periodic occurrences within a unit of time. Low-frequency waves have large wavelengths and can pass easily through clouds and over terrain following the curvature of the earth, but they can be encoded with only minimal information. High-frequency waves have short wavelengths and can be encoded with maximal information, but they have difficulty passing through clouds and over terrain, relying on line-of-sight for successful reception. In sum, low-frequency waves can carry less data over greater distances, and high-frequency waves can carry more data over shorter, line-of-sight distances. This inverse ratio between frequency and wavelength determines such things as the coverage area of a hotspot and the bandwidth of a wireless connection.

The first transmitters broadcasted a wide range of frequencies at once, which meant that a high-power transmitter could interfere with the signal of a weaker or more distant transmitter. Marconi, along with his competition, knew that if radio communication was to become viable, something had to be done about the frequency problem. The solution came in the form of tuning the transmitter and receiver to resonate at the same frequency. The tuning of signals ultimately precipitated the need for regulatory agencies, such as the U.S. Federal Communications Commission (FCC), to control the allocation of frequencies to prevent interference. However, a more immediate benefit of Marconi's solution to the interference problem, reasons John Bray (2002), "was that more than one transmitter could then be coupled to the same aerial, as could receivers to the receiving aerial" (p. 70). Each transmitter and receiver combination could be tuned to a different frequency, increasing the amount of information that could be sent at one time without interference. The difference between the highest and lowest frequency transmitted is called the bandwidth. This early ability to transmit two or more frequencies at the same time might be viewed as the beginnings of the drive for ever-increasing bandwidth. Charles Meadow (2002) explains it as follows:

> If you had some kind of transmitter capable of sending only one wave per second you could not get very much information across in that second. How much can you vary a single wave to make it a meaningful signal? The only choice would be to vary its amplitude or signal strength. Make that a two Hz and you can at least send one loud and one soft, in either order, or both loud or both soft. Jump up to 1000 Hz

(1 kilohertz or 1 kHz) and you can begin to sense how many different symbols—combinations of wave variations—you can send in a second. (p. 185)

The drive for bandwidth was further intensified by the invention of pulse-code modulation (PCM).

Invented by Alec Reeves in 1937 and granted a British patent in 1939, PCM is a means of further reducing the interference and attenuation of radio waves during the transit from one radio station to the next. PCM works by turning an analog signal into a digital signal by successively sampling the amplitude of the signal and encoding it as a binary function. In this shift from encoding information as the differences between periods of off and on states (e.g., Morse code communicates through the sequencing of long and short pulses of electricity) to the binary code of "off and on" (i.e., a digital signal), PCM increased the need for bandwidth by increasing the amount of bits per second needed to transmit the encoded message. Bray (2002) points out that "a 64,000 bit/sec telephone speech circuit requires a bandwidth, according to Hartley's Law, of at least 32 kHz, compared with the basic speech bandwidth of about 3 kHz" (p. 173). Notwithstanding a physics lesson, the inverse relationship between frequency and wavelength noted earlier is even more important to the amount of information that can be transmitted within a radio wave after PCM and the shift to binary code. High-frequency waves can be turned off and on more frequently than low-frequency waves before they can no longer be recognized as waves. Thus, more digital data can be encoded into high-frequency waves than low-frequency waves. This concept is illustrated in Figure 16.1. As the low-frequency wave is turned off and on at higher speeds, it appears to be more random and less wave-like. The higher-frequency wave retains more of its wave characteristics at the same amount of digital encoding. Sending a digitally rich message with a low frequency wave is like trying to write a novel without using vowels. It can be done, but it takes a long time and is very hard to understand.

The current drive for greater bandwidth to transmit ever-larger multimedia files means that wireless technology is being pushed further towards the visible spectrum of light and higher-energy electromagnetic waves.

Wireless communication technologies rely on radio waves to communicate over distances. As such, the space in which they can be located is much greater and more dynamic than the end of Morse's copper wire. Radio transmitters and receivers can be placed just about anywhere (communications satellites orbit the Earth) and still communicate, so long as the strength and frequency of the radio wave are taken into account. In this

Figure 16.1. High-and Low-Frequency Waves with Digital Data Encoded

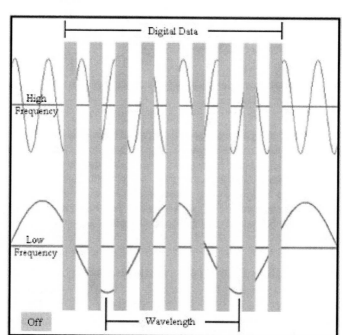

way, radio communication is integral to wireless and mobile technologies. Radio waves are the part of the equation that keeps users connected when they are mobile. However, the questions still remain as to why users need to be mobile in the first place, and what exactly is being mobilized.

MOBILE: A BRIEF HISTORY

Mobile, etymologically stemming from the Latin *mobilis*, means tending to move. In terms of the history of mobile technologies, what was being moved technologically was the ability to perform computations, not the ability to communicate. This is a key factor that once separated "mobile" from "wireless" in wireless and mobile technologies, with mobile and computation on one side of the ratio and wireless and communication on the other side. In this sense of the ability to perform computations while moving or tending to move, the term "mobile" has a long history that can be reviewed quickly.

Mechanical Computational Aids

The history of mobile computation arguably begins with the human brain. John Kopplin (2002) contends

> The first computers were people! That is, electronic computers (and the earlier mechanical computers) were given this name because they performed the work that had previously been assigned to people. "Computer" was originally a job title: it was used to describe those human beings (predominantly women) whose job it was to perform the repetitive calculations required to compute such things as navigational tables, tide charts, and planetary positions for astronomical almanacs. (¶ 1)

The first mechanical computers were not the automated calculators that we think of today, but mechanical devices to aid humans—Kopplin's human computers—in completing long and difficult computations. The abacus is the first of these mechanical aids: The oldest surviving abacus dates to Babylonia c. 300 BCE, whereas earlier versions probably consisted of various ways of making lines and marks in dust on the ground (Kopplin, 2002, ¶ 2). The abacus, in its many forms and iterations, was used by every advanced civilization at one time or another.

As aids to the human computers, other mechanical computers were also small and uncomplicated to use. Napier's Bones (the next most important mechanical computer in the history of mobile computing) was invented by the Scotsman John Napier and consisted of the logarithms he invented in 1614 carved onto a series of ivory rods. The rods were placed side by side according to the numbers to be multiplied and the result was computed by adding the numbers down and across the logarithmic grid. Napier's Bones later led to the development of the slide rule, through some refinement by Edmund Gunter and then by William Oughtred, for use in astronomy and navigation. Not until the mid-20th century was the slide rule found to be in common use, having gone through a number of further refinements. Michael Williams (1990) observes that at the slide rule's apex was Fuller's Slide Rule, a version of the slide rule that "was equivalent to a standard slide rule over eighty-four feet long, yet could be easily held in the hand" (p. 33). In fact, all of the above computers (the early mechanical and human ones) are mobile. They can be moved around with ease, and with some determination, a user can perform computations while moving.

An important note here is that the term "portable," although currently used to designate various types of laptop computers (e.g., the ultra-portable laptop), is etymologically different from the term mobile as I am using it here. Portable, etymologically stemming from the Latin *portabilis*, means able to be carried. This term is in contrast to mobile, which I defined above as tending to move. A good distinction between portable and mobile is that portable suggests that you can move your computer from place to place and mobile suggests that you can continue to compute while you are doing so.

Computational Machinery

During the refinement of Napier's Bones into the slide rule and to the slide rule's heyday, computational machines were being developed as well. Computational machines are mechanical computers that rely on gears and rods to perform computations automatically. Rather than aiding humans with the calculation, computational machinery performs the calculation for humans. Beginning with Wilhelm Schickard's Calculating Clock, they are as follows:

- Wilhelm Schickard's "Calculating Clock" (ca. 1623)
- Blaise Pascal's "Pascaline" (1642)
- Gottfried Wilhelm Leibniz's "Stepped Reckoner" (1674)
- Charles Xavier Thomas de Colmar's "Arithmometer" (1820)
- Frank S. Baldwin's "Baldwin's 1875 Machine" (1873)
- Dorr E. Felt's "Comptometer" (1887)
- Otto Steiger's "Millionaire" (ca. 1890)[2]

Most of the computational machines were rather heavy and difficult to move. Although they may have been portable with some effort, they certainly were not readily mobile. Even the smaller machines, such as the Pascaline and the Arithmometer, were prone to error if knocked around or jarred, and the Comptometer, which was roughly the size of a shoebox, required both hands to operate when computing large numbers. However, these computational machines are important to note because they are the forerunners of the modern computer, connecting Joseph Marie Jacquard's power loom and the invention of the punch card to Charles Babbage's "Analytic Engine" to Hermann Hollerith's "Hollerith desk" and ultimately to the founding of the Tabulating Machine Company, which becomes IBM.[3]

Mobile Computers

IBM's role in the development of the computer industry is well document-ed, but its role in the development of mobile computing is just as impor-tant if not as well documented. Not until IBM produces the ThinkPad in 1992 did the laptop industry take off, with IBM selling 20 million ThinkPads between 1992 and 2004. However, IBM did not invent the mobile comput-er with the development of the ThinkPad. This distinction goes to Epson. Christopher Null (2005) reports that

> Epson's HX-20, introduced in 1982, was the first computer described as a "laptop." These were tiny machines designed to be propped in your lap instead of used on a desk. The HX-20 tipped the scales at barely 3 pounds, and it included a built-in tape drive and a tiny print-er. Best of all, unlike its bigger forebears, this machine could run on batteries: The HX-20 had an impressive 50 hours of life on its recharge-able nickel-cadmium cells. (Section 3, ¶ 1)

The criterion to note is not the moniker "laptop" but the fact that the HX-20 ran on batteries making this the first truly mobile computer. Earlier portable computers, such as the IBM 5100 Portable Computer, the Osborne Computer, and the Xerox Notetaker, weighed upwards of 40 pounds and had to be plugged in for power. As such, they are not mobile computers.

One exception to this history is the development of the MOBIDIC (Mobile Digital Computer) by the Sylvania Electric Products Company for the U.S. Army Signal Corp in 1957. Housed in an Army trailer and trans-ported by truck (see Figure 16.2), the MOBIDIC was designed to coordinate the collection and distribution of intelligence from smaller portable com-puters transported by jeep around the battlefield. Because the MOBIDIC carried both its power source and transportation source with it and could operate and move at the same time, the MOBIDIC qualifies as the first truly modern mobile computer.

Figure 16.2. Mobidic Cutaway View[4]

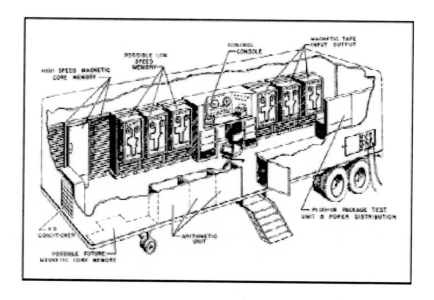

WIRELESS AND MOBILE: QUESTIONING THE TERMS

The common factor that connects *wireless* and *mobile* as they pertain to wireless and mobile technologies is the liberation from wires. The wires must be replaced or unplugged. The wireless modem has become a component of the mobile computer, and thus, we find ourselves conflating *wireless* and *mobile* to *wireless and mobile*.

Wireless Is to Communication as Mobile Is to Composition

The separate histories of the terms *wireless* and *mobile* created a proportion of two ratios: wireless is to communication as mobile is to computation. In this proportion, the emphasis was placed on the first term in each ratio (that is, wireless and mobile). The conflation of these terms into *wire-*

less and mobile changes that proportion into an equation: wireless and mobile equals communication and computation. The current conflation of the terms places the emphasis on the left side of the equation (again, wireless and mobile). As a teacher and researcher of composition, I wish to appropriate the equation and make a new proportion, one that includes a place for composition. In my new proportion, the term *composition* replaces the term *computation* in the right hand ratio, and the new proportion becomes wireless is to communication as mobile is to composition. I also shift emphasis from the first terms on either side of the ratio to the second terms on either side of the ratio (that is, communication and composition) because I believe that composition is about communicating through writing. My shift in emphasis is illustrated in Figure 16.3. Appropriating the terms and the proportion created by them helps me question the currently held equation and its exclusion of composition.

Throughout the history of telegraphy, both before and after Morse's Telegraph, the great limitation to communication has been bandwidth (used loosely when applied to ancient modes of telegraphy). When people used bonfires to communicate across the plains, the message communicated by the bonfire had to be decided upon in advance. Once this message

**Figure 16.3. Shifting Emphasis from Wireless and Mobile
to Communication and Composition**

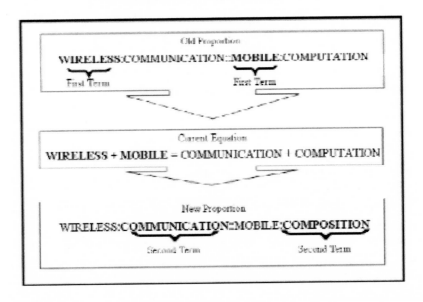

was chosen, no other message could be sent. The fire was either burning or not. This was a very narrow bandwidth indeed. Smoke signals and drum beats can communicate more messages, but the message is easily intercepted or interfered with. Horse and rider could only carry so much weight if the message was to be delivered in a timely manner.

In other words, the very mode of communication limits what can be communicated. This same problem is exhibited by the telegraph as well. I have already discussed the limitations of transmitting a message via the copper wire, but I have not discussed how the message can be altered to mitigate these limitations. Here, then, is a critical gap where compositionists can assert themselves. It is our work as rhetoricians and compositionists to attend to the possibilities and constraints of language, to prepare students to understand those potentials and limitations, and to work as active agents. Consider that because of the limited bandwidth of the telegraph, messages were priced by the word. To save money, a long message would be shortened as much as possible. If I had been alive during the last half of the nineteenth century, I might have been inclined to conduct an ethnographic study of the ways in that people shortened messages to reduce the cost. I might then have developed a sort of shorthand which users could employ to encode a long message in the fewest words possible. This shortened, coded message could then be taken to the telegraph office for transmission. The same is true for Chappé's system of semaphore giants. Rather than relying on the positioning of arms for each symbol in the code, I might have developed a code similar to American Sign Language, in which some words are signed and others are spelled out. The same is true for the modes of communication from wired to wireless; compositionists are experts in the structure of languages and even more expert in the rhetorical context—economic and social—to help solve bandwidth problems in many forms of communication.

Throughout the history of mobile computing, the great limitation to computation has been the ability to store the numbers as the calculations were performed. As I noted earlier, the first abacus probably relied on the making of lines and marks in the dust. At this early stage, computation and composition might well be considered the same thing. Etymologically, computation and composition are not so far apart. Computation, stemming from Latin, means the act of putting together (*com*) to make clear (*putare*), whereas composition, also stemming from Latin, means the act of putting (*ponere*) together (*com*). From the perspective of composition, computers are used for composing, regardless of media or multimedia.

From the composition perspective, how might the mechanical computational aids and the computational machinery described here have aided the human *composer* rather than the human computer? The same punch card that Jacquard used to automate his loom was later used by Hollerith to take the 1890 census.[5] Jacquard was able to create intricate and beautiful tapestries using punch cards that stored a pattern for the loom to follow. Might I have written messages to the world with those punch cards, the loom then translating them into giant documents to be hung about the city like billboards?

These are, to be sure, limited examples of how a compositionist might revise the histories which conflated *wireless* and *mobile* into *wireless and mobile* in order to recontextualize the technology and emphasize the roles of communication and composition, but this revision must be conducted because the age of integration is upon us. Manufactures of wireless and mobile technologies are striving to make the one device that does it all. Cell phones now play music and take digital pictures. Music players now play video and download both music and videos wirelessly. Mobile computers, such as PDAs, have merged with cell phones. These devices provide all the capabilities of a cell phone, a music player, and a computer, with a large footprint for wireless access as well as running Microsoft WINDOWS applications—truly a *wireless and mobile* technology. The slick and sleek design of these devices suggests that manufacturers are targeting the young adult market—a market that is already bringing these devices into our classroom and changing our pedagogies. The way for us to assert critical agency is to recontextualize the technologies and build a new set of relations that include composition pedagogies and research methods.

CONCLUSION

The trajectory of wireless and mobile technologies is not that of a bullet. While it may seem to have as much energy and momentum as a bullet leaving a gun, we cannot expect that the trajectory of wireless and mobile technologies will arc upwards and eventually fall. My personal predictions for wireless and mobile technologies are dark. Soon students and teachers won't meet in classrooms at all. Face to face (f2f) teaching will become a thing of the past, and learning will be something that takes place on the bus to and from anywhere but school.

We live in a wireless and mobile culture, but we often take the ramifications of this for granted. As teachers and scholars of composition, we need to question this in our lives and in our pedagogies. What follows is a list of terms pertaining to wireless and mobile technologies relevant to teachers and scholars of composition. I examine how these various terms might shape composition pedagogies and provide my analysis after the definition. In this way, I hope to suggest ways composition teachers and scholars might engage these technologies critically. As the chapters in this collection demonstrate, there are both gains and losses when wireless and mobile technologies are moved into the classroom and the university. This history has not been written yet. Nonetheless, we will have to work hard to make the terms of that history our own.

KEY TERMS FOR GOING WIRELESS

Asynchronous: With respect to communication, a relationship in which a transmission occurs without a regular or predictable time specified.[6]

As discussed in this collection, the use of mobile computers in the service of ethnography provides the researcher with valuable tools for note taking and analysis of data. However, without wireless capabilities, the recording of data and the analysis of that data will be asynchronous. A wireless connection would enable a researcher to send data back to an assistant for simultaneous analysis.

Bluetooth: A short-range radio technology developed and marketed by the Bluetooth Special Interest Group to facilitate transfer of information among computers and other devices and the Internet.

Bluetooth was designed without support for Internet Protocols, so its market is currently growing in the area of wearable computers, where it is the means of wirelessly connecting the computer to peripherals of all types. Bluetooth also comes standard on many PDAs, where it is used to transfer data between a PDA and desktop computer. Bluetooth is not compatible with the Ethernet or Wi-Fi.

Ethernet: Developed in 1976, this LAN architecture was the basis for the IEEE 802.3 standard and is often used synonymously. Ethernet is a widely used, wired standard for use with coaxial cables, Ethernet cables, and fiber optics, and supports transfer rates of 10 Mbps. Newer versions of the standard, known as Fast Ethernet and Gigabit Ethernet, can support transfer rates of 100 Mbps and 1000 Mbps, respectively.

> Ethernet is the standard with which most wired classrooms and wired homes are wired. The Ethernet data port resembles an over-sized telephone jack and comes standard on new mobile computers. Because Ethernet is common, more students and teachers are comfortable plugging in and trouble shooting their connections. However, once a computer is plugged into the Ethernet, it is no longer wireless and no longer mobile.

Firewall: A computer or program that regulates traffic between an external network and a private network in order to protect the private network from the external network.

> In this case the private network may be a single computer. Because firewalls run in the background, they do not draw attention to themselves until there is an attack or a problem. In addition, the marketing of computer security relies on scare tactics. Combining the two together means that most students and teachers will not know how to trouble shoot or reconfigure a firewall when a problem occurs and will be hesitant to do so. This is doubly true when the firewall is preventing access to a wireless network.

Hotspot: The area in which the radio signals from a public, wireless LAN can be received by a compatible wireless transceiver.

> Most hotspots are located in high traffic areas, where people naturally congregate. For the teacher experimenting with the mobile classroom, the hotspot is necessary and problematic. On the one hand, the hotspot allows students and teacher to roam away from the classroom while remaining connected to each other over the wireless LAN. On the other hand, the hotspot allows students and teachers to congregate loosely, spreading out from one another while remaining in contact with one another over the wireless LAN. Student and teacher might be so spread out that the one no longer knows what the other is up to. This could be just the freedom a shy student needs to open up or just the freedom a precocious student needs to open up a beer.

IEEE 802.11: The Institute of Electrical and Electronics Engineers family of standards for wireless LANs.

An important factor to note here is that the standards are embedded in the hardware that transmits the signal. As such, wireless computers using different standards may not be compatible with one another or the wireless network. For instance, IEEE 802.11a is not compatible with IEEE 802.11b and g, although IEEE 802.11b and IEEE 802.11g are compatible with each other and can be transmitted by the same hardware. Another important factor to note is that, in general as the letter designation of the IEEE 802.11 standards increases, so too does the range and bandwidth of the signal. The exception to this comes with the IEEE 802.11a standard, which provides up to 43 Mbps more than IEEE 802.11b. IEEE 802.11a transmits at the 5 GHz band, providing a higher quality of transmission but a shorter range, and it is essentially no longer marketed. Either way, administrators will need to specify the standard when requiring students to provide their own wireless laptops.

Instant Messaging (IM): Synchronous, written communication over the Internet with an individual via a private chat room.

Resembling the telegraph in the manner by which users truncate the message into as few as possible typed characters, this mode of communication is ripe for ethnographic study. Stories that describe the infiltration of IM text into the academic essay are already circulating. This infiltration will only increase if teachers of composition begin using IM in their pedagogies uncritically.

Local Area Network (LAN): A network that connects a set of computers common to a geographic location, usually within a structure or related structures.

Most LANs connect the computers on a network to data and peripherals, such as printers, elsewhere on the LAN. For the wireless and mobile student, this means that a document can be printed immediately after drafting it regardless of where it was drafted or where the printer is located.

Media Access Control (MAC) Address: The unique identification number of each computer on a network promulgated by the modem.

MAC addresses are extremely important to wireless and mobile technologies because they are the identification that the firewall checks to determine whether a particular computer is allowed access to a network. The MAC address is a hardware address and cannot be transferred to another computer with removing and reinstalling the hardware, tasks the average student probably is not familiar with. As such, if a student borrows a friend's computer, the MAC address will not be recognized, and the student will not be allowed onto the network.

Multiple-In, Multiple-Out (MIMO): A multiplexing system in which data is sent and received over more than one antenna to increase range and bandwidth.

Current MIMO technology is compatible with both IEEE 802.11b and g. However, MIMO technology increases the transmission of data to ranges approaching one-half mile. This range probably allows more freedom than any student needs, but truly emphasizes the mobile aspect of wireless and mobile technology while maintaining the wireless aspect.

Short Messaging System (SMS): A mobile telephone service that allows a user to send and receive short messages (≤160 characters) synchronously on the same radio wave as voice transmissions.

Similar in effect to Instant Messaging, the limited number of characters that can be sent at one time requires users to truncate their spelling and choose words specifically to shorten the length of the message, resulting in a creole that arises from the contact of language and technology as opposed to the contact of two languages.

Synchronous: With respect to communication, a relationship in which a transmission occurs simultaneously or nearly simultaneous.

Chat rooms are synchronous because everyone in the chat room can send their messages at the same time. Protocols must be developed and enforced to ensure that students who type slowly or poorly or students who are on a slow connection are not forced out of the conversation by faster typists or students who are on a faster connection. Furthermore, this synchronous technology has influenced the perception of other technologies such as email, causing students to expect an immediate reply to their queries.

Wireless Fidelity (Wi-Fi): This generic term, made popular by the Wi-Fi Alliance, refers to a network running any of the IEEE 802.11 wireless standards.

As noted above in the entry on IEEE 802.11, not all Wi-Fi is compatible. Administrators will have to choose a standard and notify students well in advance if students are required to come prepared with the proper hardware.

ENDNOTES

1. Invented by Claude Chappé during the French Revolution to relay messages, this semaphore system consisted of movable arms atop a support tower set on hilltops every ten kilometers or so. Ropes were used to manipulate the position of the arms to indicate a number of predetermined codes for letters of the alphabet. Charles Meadow (2002) notes that a message could be sent over 700 kilometers in about 40 minutes (p. 89).
2. For more information on these computational machines, see William Aspray (1990), *Computing before computers.*
3. For more information on the chain of events that lead to the invention of the computer, see Georges Ifrah (2000), *The universal history of computing: From the abacus to the quantum computer.*
4. This image (retrieved from *Wikipedia, the free encyclopedia* at http://en.wikipedia.org/wiki/Image:MOBIDIC_cutaway.jpg) is a work of a U.S. Army soldier or employee, taken or made during the course of the person's official duties. As a work of the U.S. federal government, the image is in the public domain.
5. For a detailed analysis of how Jacquard's punch card led to the computer revolution, see James Essinger (2004), *Jacquard's web: How a hand-loom led to the birth of the information age.*
6. For definitions more technical than I provide here, see Julie K. Pertersen (2002), *The telecommunications illustrated dictionary.*

REFERENCES

Aspray, William. (Ed.). (1990). *Computing before computers*. Ames: Iowa State University Press.

Bray, John. (2002). *Innovation and the communications revolution: From the Victorian pioneers to broadband internet*. London: Institution of Electrical Engineers.

Burke, Kenneth. (1969). *Language as symbolic action: Essays on life, literature, and method*. Berkeley: University of California Press.

Burke, Kenneth. (1973). *Philosophy of literary form* (3rd ed.). Berkeley: University of California Press. (Original work published 1941)

Essinger, James. (2004). *Jacquard's web: How a hand-loom led to the birth of the information age*. Oxford, England: Oxford University Press.

Foucault, Michel. (1972). *The archaeology of knowledge* (A. M. Sheridan Smith, Trans.). New York: Pantheon.

Ifrah, Georges. (2000). *The universal history of computing: From the abacus to the quantum computer*. New York: John Wiley.

Kopplin, John. (2002). An illustrated history of computers: Part one. Retrieved March 23, 2003, from http://www.computersciencelab.com/computerhistory/history.htm

Meadow, Charles. (2002). *Making connections: Communication through the ages*. London: Scarecrow Press.

Null, Christopher. (2005). The birth of the notebook [Electronic version]. *Mobile Magazine: Technology to Go, 3*, 1-11. Retrieved July 24, 2005, from http//:www.mobilemagazine.com/archives/2005/03/the_birth_of_1.html

Petersen, Julie K. (2002). *The telecommunications illustrated dictionary* (2nd ed.). New York: CRC Press. (Original work published 1999)

Williams, Michael R. (1990). Early computing. In William Aspray (Ed.), *Computing before computers* (pp. 3-58). Ames: Iowa State University Press.

ABOUT THE CONTRIBUTORS

John F. Barber, Ph.D. co-directs (with Dene Grigar) the Digital Technology and Culture Department at Washington State University Vancouver. His research and publication examines potentials wrought by shifting relationships between technology, art, science, and the humanities. For example, *New Worlds, New Words: Exploring Pathways for Writing about and in Electronic Environments*, a volume edited with Dene Grigar, focuses on the future of writing resulting from its move to electronic spaces. Barber has published chapters in *Transdisciplinary Digital Art: Sound, Vision and the New Screen, Texts and Technology, The Online Writing Classroom, Electronic Networks, HighWired*, and *Studies in Technical Communication*; articles in print journals like *Readerly Writerly Text, Works and Days*, and *Pre/Text*; and articles in electronic journals like *Leonardo, Fine Art Forum*, and *Kairos*. He is developer and curator of *Brautigan Bibliography and Archive*, an information portal focusing on writer Richard Brautigan. An upshot of this project is *Richard Brautigan: Essays on the Writings and Life*, as well as entries regarding Brautigan in the *Encyclopedia of Beat Literature*.

Olin Bjork is a Marion L. Brittain postdoctoral fellow at the Georgia Institute of Technology, where he teaches courses in composition, literary/cultural studies, and technical communication. His research centers on John Milton, textual studies, the digital humanities, and instructional technology. He recently received his Ph.D. in English from the University of Texas at Austin, where he served as assistant director of the Computer Writing and Research Lab (CWRL), taught literature and composition courses in the CWRL's computer-assisted classrooms, and worked as the English department's webmaster. For the past several years, he has been developing with John

Rumrich a digital "audiotext" or classroom edition of Milton's *Paradise Lost* (http://www.laits.utexas.edu/miltonpl).

Kevin Brooks is Associate Professor of English and Writing Program Administrator at North Dakota State University. He has published on technology and writing instruction in *Pedagogy, Into the Blogosphere*, and *Kairos* and is working on a book project tentatively titled *McLuhan for Compositionists*.

Nicole R. Brown is an Assistant Professor of English at Western Washington University. Her areas of specialization and interests include rhetoric and composition, technical writing, cybercultural studies and community writing/learning. She has conducted research related to online community in social and academic contexts and has presented scholarly papers and published articles on computers and writing, workplace writing, and visual rhetoric.

Donna Daulton recently earned her MA in linguistics at the University of Memphis, and has taught K-3 special education and adult ESL, and, has designed and led technology workshops for English instructors. Her current work investigates the oral-literate continuum as expressed in fairytales, and student technology use as related to personality types.

Teddi Fishman serves as the director of the Center for Academic Integrity in association with the Rutland Center for Ethics where she is currently researching issues of ethics and technology. She continues her association with the Masters of Professional Communication (MAPC) faculty at Clemson University and her work with the Pearce Center for Communication, teaching advanced (upper level undergraduate as well as graduate) writing courses in "wired classrooms" and online spaces. She is currently collaborating on another book chapter on the subject of ethics in online communication in professional settings.

Simson L. Garfinkel is an Associate Professor at the Naval Postgraduate School in Monterey, California, and a fellow at the Center for Research on Computation and Society at Harvard University. His research interests include computer forensics, the emerging field of usability and security, personal information management, privacy, information policy and terrorism. Garfinkel is the author or co-author of fourteen books on computing. He is perhaps best known for his book *Database Nation: The Death of Privacy in the 21st Century*. Garfinkel's most successful book, *Practical UNIX and Internet Security* (co-authored with Gene Spafford), has sold more than

250,000 copies and been translated into more than a dozen languages since the first edition was published in 1991. Simson Garfinkel received three Bachelor of Science degrees from MIT in 1987, a Master's of Science in Journalism from Columbia University in 1988, and a PhD in Computer Science from MIT in 2005.

Dene Grigar is an Associate Professor and Director of the Digital Technology and Culture program at Washington State University Vancouver. Her books include *New Worlds, New Words: Exploring Pathways in and Around Electronic Environments* (with John Barber) and *Defiance and Decorum: Women, Public Rhetoric, and Activism* (with Laura Gray and Katherine Robinson); media art works include "Fallow Field: A Story in Two Parts" and "The Jungfrau Tapes: A Conversation with Diana Slattery about *The Glide Project*," both of which appeared in *Iowa Review Web* in October 2004, and *When Ghosts Will Die* (with Canadian multimedia artist Steve Gibson), a piece that experiments with motion tracking technology to produce narrative. Her most recent work, also with Gibson, is the *MINDful Play Environment*, a live, interactive game environment.

Will Hochman is s Professor of English and Technology Coordinator at Southern Connecticut State University. He has published scholarly work on hypertext, J.D. Salinger, writing pedagogy and writing program administration, and Hochman's latest book is *Freer*, a collection of poems that was published by Pecan Grove Press. He is also the poetry editor of *War, Literature & the Arts*.

Johndan Johnson-Eilola works as a Professor in the Department of Communication & Media at Clarkson University. His most recent books include *Datacloud: Toward a New Theory of Online Work* (Hampton Press, 2005) *Central Works in Technical Communication* (co-editor with Stuart A. Selber, Oxford 2004), and *Writing New Media* (with Anne Wysocki, Cynthia Selfe, and Geoff Sirc, Utah State University Press, 2004).

Joseph G. Jones is an Assistant Professor in Composition and Professional Writing and has directed the First-Year Composition program at the University of Memphis. His primary research interests engage composition theory and pedagogy, writing assessment, and the histories of high school and college English studies.

Loel Kim is an Associate Professor in the Professional Writing program at The University of Memphis and former director of the Center for the Study of Rhetoric and Applied Communication (CSRAC), an interdisciplinary group of scholars from the departments of English and Communication. Her recent research centers on the social and cognitive factors involved in user perceptions of human agents in interactive systems and online communities.

Amy C. Kimme Hea is Associate Director of the Writing Program and Associate Professor in the Rhetoric, Composition, and Teaching of English program at University of Arizona. Her research interests include web and wireless teaching and learning, teacher development, and professional writing theory and practice. Her work appears in the anthologies and journals including *Computers and Composition, Kairos, Educare/Educare,* and *Reflections: A Journal of Writing, Service-Learning, and Community Literacy.*

Karla Saari Kitalong teaches in the Department of English at the University of Central Florida in Orlando, where she is associated with both the Texts and Technology Ph.D. program and the B.A. and M.A. programs in technical communication. Her research interests include visual representation and usability in/of technological contexts. She has a long-standing interest in how media representations connect to identity formation and to the acquisition of technological literacies.

Beth Martin holds an M.A. with an emphasis in technical writing and an M.L.I.S. Her M.A. research used genre theory and embodied knowledge to examine how genre inhibits or spurs innovation. Her M.L.I.S. research focused on information seeking behaviors. Prior to academia she worked for 15 years as a network engineer and now serves as a communication consultant for a major financial institution.

Lisa Meloncon has a PhD in Rhetoric and Composition with a specialization in technical communication from the University of South Carolina. Lisa teaches at the University of Cincinnati in the Professional Writing Program. Her research works at the center of an interdisciplinary nexus: technical communication, rhetoric, technology, history, and visual literacy. She also has over 15 years of technical writing industry experience.

David Menchaca is the Director of the Professional Writing Program at Washington State University Vancouver and cofounder of the Learning Games Initiative (with Ken McAllister). His primary research interest is the rhetoric of technology with secondary interests in information design, games studies, and classical rhetorics.

Ryan "Rylish" Moeller is an Assistant Professor in the English Department at Utah State University. He teaches courses in technical writing, rhetorical theory, and the rhetorics of technology. His research is focused on the relationships among technique, technology, and rhetorical agency. His work has appeared in *Technical Communication Quarterly, Kairos, Works and Days*, and in book chapters. He is currently working on a book manuscript that examines the rhetoric of consumer electronics.

Mike Palmquist is Professor of English and University Distinguished Teaching Scholar at Colorado State University, where he directs the Institute for Learning and Teaching. His scholarly interests include writing across the curriculum, the effects of computer and network technologies on writing instruction, and new approaches to scholarly publishing.

Mya Poe, Ph.D. is Director of Technical Communication at the Massachusetts Institute of Technology. At MIT, she teaches writing in the departments of Electrical Engineering and Computer Science and the Harvard-MIT graduate program in Health Sciences Technology. Dr. Poe's research interests include visual communication in the sciences, writing across the curriculum, and writing assessment.

Susan L. Popham is an Associate Professor in the Professional Writing and Composition programs and former director of the first-year composition program at the University of Memphis. She conducts research in writing program administration and is finishing a book on medical business writing.

John Pedro Schwartz is Assistant Professor of English at American University of Beirut, where he teaches modern British and American literature and critical theory. He is completing a manuscript that applies the insights of museum studies to the analysis of literary texts. He has published an essay in the *JJQ* on monument discourse in *Finnegans Wake*. An article on teaching multiliteracies through the museum will appear in the September 2008 issue of *College English*. He has taught composition and literature courses in computer-assisted classrooms and incorporated MOO, Web, blog, podcasting, and other new media into his pedagogy.

Stuart A. Selber is an Associate Professor of English and an affiliate Associate Professor of Information Sciences and Technology at Penn State. He is the author of *Multiliteracies for a Digital Age* (SIUP 2004) and the co-editor, with Johndan Johnson-Eilola, of *Central Works in Technical Communication* (Oxford UP, 2004). Both books won awards from the National Council of Teachers of English.

Clay Spinuzzi is an Associate Professor of rhetoric at The University of Texas at Austin. Spinuzzi's interests include research methods and methodology, workplace research, and computer-mediated activity. His book *Tracing Genres through Organizations,* published in 2003, was named NCTE's 2004 Best Book in Technical or Scientific Communication. His second book, *Network,* was published in 2008.

Emily A. Thrush is a Professor in the Professional Writing and Applied Linguistics programs at the University of Memphis. She has been an Academic Specialist for the USIA/State Department in several countries around the world, and served as a Senior Fulbright Fellow in Mexico in 2000-2001. Her research interests include information design for the World Wide Web, the impact of computers on writing, and multicultural issues in technical communication. She has supervised the English Department computer labs for 15 years.

Melinda Turnley is an Assistant Professor of Writing, Rhetoric, and Discourse at DePaul University. Her research and teaching interests highlight issues of pedagogy, teacher and program development, and critical approaches to technology. Her work has appeared in *Computers and Composition, Kairos,* and *Rhetoric Review.*

Kathleen Blake Yancey is Kellogg W. Hunt Professor of English at Florida State University, where she directs the graduate program in Rhetoric and Composition Studies. Past President of the Council of Writing Program Administrators and Past Chair of the Conference on College Composition and Communication, she is President of the National Council of Teachers of English. She co-founded and co-directs the International Coalition on Electronic Portfolio Research, a consortium of over 45 institutions documenting the learning that takes place inside and around electronic portfolios. The author of over 60 chapters and refereed articles, she has edited or authored 10 books, including *Portfolios in the Writing Classroom* (1992), *Assessing Writing across the Curriculum* (1997), *Electronic Portfolios* (2001), and *Delivering College Composition; The Fifth Canon* (2006), the last of which won the 2006-2007 WPA Best Book Award. Her current projects include the volume based on her CCCC Chair's Address, *Composition in a New Key: A Theory and Practice of Composition in the 21st Century.*

AUTHOR INDEX

SUBJECT INDEX

LaVergne, TN USA
09 June 2010
185432LV00002B/9/P